Signal Processing Techniques for Communication

The reference text discusses signal processing tools and techniques used for the design, testing, and deployment of communication systems. It further explores software simulation and modeling tools like MATLAB, GNU Octave, Mathematica, and Python for modeling, simulation, and detailed analysis leading to comprehensive insights into communication systems. The book explains topics such as source coding, pulse demodulation systems, and the principle of sampling and aliasing.

This book:

- Discusses modern techniques including analog and digital filter design, and modulation principles including quadrature amplitude modulation, and differential phase shift keying
- Covers filter design using MATLAB, system simulation using Simulink, signal processing toolbox, linear time-invariant systems, and non-linear time-variant systems
- Explains important pulse keying techniques including Gaussian minimum shift keying and quadrature phase shift keying
- Presents signal processing tools and techniques for communication systems design, modeling, simulation, and deployment
- Illustrates topics such as software-defined radio (SDR) systems, spectrum sensing, and automated modulation sensing

The text is primarily written for senior undergraduates, graduate students, and academic researchers in the fields of electrical engineering, electronics and communication engineering, computer science, and engineering.

Signal Processing Techniques for Communication

K. C. Raveendranathan

CRC Press
Taylor & Francis Group
Boca Raton London New York

CRC Press is an imprint of the
Taylor & Francis Group, an **informa** business

Designed cover image: K. C. Raveendranathan

First edition published 2025
by CRC Press
2385 NW Executive Center Drive, Suite 320, Boca Raton FL 33431

and by CRC Press
4 Park Square, Milton Park, Abingdon, Oxon, OX14 4RN

CRC Press is an imprint of Taylor & Francis Group, LLC

ISBN: 978-1-032-75649-3 (hbk)
ISBN: 978-1-032-86453-2 (pbk)
ISBN: 978-1-003-52758-9 (ebk)

DOI: 10.1201/9781003527589

Typeset in Nimbus font
by KnowledgeWorks Global Ltd.

To
My wife Sobha and daughters Anuja and Aabha

Contents

Forward

The advancements in signal processing (DSP) techniques significantly influence the evolution of communication systems. As we progress further into the digital information and networking age, from 5G to 6G networking, the importance of the DSP techniques becomes ever more paramount.

This book, *"Signal Processing Techniques for Communication Systems,"* written by an eminent academician and a close professional colleague of mine, Prof. K. C. Raveendranathan, provides a comprehensive exploration of the fundamental and advanced methods that underlie modern DSP and communication systems.

The book is meticulously crafted to serve as a textbook for senior undergraduate and graduate-level courses, with a clear focus on bridging the gap between theoretical concepts and practical applications. It takes the reader on a journey from the basic principles of signals and systems to the complexities of digital signal processing, all while maintaining a strong emphasis on the practical tools and software that are essential in the field today.

What sets this book apart is its integration of software tools such as MATLAB, GNU Octave, Python, and Mathematica, which are not merely supplementary but are woven into the very fabric of the learning process. By doing so, the author ensures that readers gain hands-on experience in modeling, simulation, and analysis, which are critical for developing a deep understanding of communication systems. The inclusion of source codes and detailed programming instructions further enhances the practical value of this book.

The text is structured to facilitate a logical progression of ideas, beginning with an introduction to the fundamental concepts of signals and systems. In early parts of the book, the reader is introduced to various signal operations and transformations, providing the necessary foundation for understanding more complex topics. The latter half of the book is dedicated to more specialized topics in communication systems such as various modulation schemes, noise modeling, and mitigation techniques, providing a comprehensive overview of the challenges and solutions in modern communication systems. Finally, the book concludes with a look at advanced topics in wireless communication and machine learning for communication systems, reflecting the cutting-edge research and trends in the field.

In summary, "Signal Processing Techniques for Communication Systems" is an authoritative and comprehensive resource that stands out for its practical approach and in-depth coverage of both fundamental and advanced topics. I am sure this book is a valuable asset to students, professionals, and academicians.

Prof. B. S. Manoj, PhD, FNAE, FIETE, SMIEEE
Board member, Membership Services Board, IEEE Communications Society
Professor and Associate Dean
Indian Institute of Space science and Technology (IIST), Trivandrum 695547, India
Email: bsmanoj@ieee.org, URL: https://www.iist.ac.in/avionics/bsmanoj

Preface

Signal Processing techniques, especially in the digital domain, are gaining a wide significance in communication engineering. This book aims to introduce various signal-processing tools that would play a pivotal role in the design, development, and deployment of communication systems. This book is intended as a textbook on Signal Processing Techniques for Communication Systems for a one-semester, senior undergraduate or graduate level course in Electronics and Communication Engineering.

MATLAB® is the *language* of the technical computing fraternity and several courses like the Signals and Systems, Communication Theory, Control Systems, Analog and Digital Communication Systems, and Digital Signal Processing can be taught more effectively and easily using this high level-programming language. MATLAB is a highly user-friendly and can be learned quite fast.

In Chapter 1, we begin with a study of fundamental concepts of systems and signals and various transforms of signals and progress to the application of Digital Signal Processing tools in the design of communication subsystems. In Chapter 2, we explore software simulation and modeling tools like MATLAB, GNU Octave, Python, and Mathematica for the modeling, simulation and detailed analysis leading to comprehensive insights in Communication Systems. MATLAB and GNU Octave, as well as Mathematica are widely used software tools for the signal processing associated to communication systems. Python, on the otherhand, is the most widely used and popular programming language as reported by the IEEE. We will introduce the basics of these software tools giving short introduction to the associated programming structures. We will also give detailed pointers to programming using these softwares. Source code of a multitude of software simulations form an integral part of this book.

In the third chapter, we explore signal processing being done in frequency domain. Various transforms like Laplace transform, Fourier transform, z-transform, Discrete Cosine Transform (DCT), and Hilbert transform and their inverse transforms are discussed. The concept of convolution is introduced. The process of obtaining the output of an NLTV system is also brought out.

In Chapter 4, we discuss analog and digital modulation schemes like AM, AM-DSB, SSB, VSB, and pulse modulation schemes like PSK, QPSK, QAM, MSK, and GMSK, and its several variants, MIMO, TDMA, FDMA, CDMA, OFDM and other modern communication paradigms. Noise analysis, and modeling of noise in communication systems is a highly significant area in the study of signal theory and communication systems. We will extensively study the various noise models such as AWGN, Salt & Pepper noise, ISI, CCI, and various concepts associated with design of systems that can result in noise mitigation.

In Chapter 5, the design of analog subsystems is dealt with. The channel noise and various filters like LPF, HPF, BPF, BRF (notch filter) are discussed in detail.

The design of analog filters using appropriate MATLAB tool kit is discussed. Digital filter design using MATLAB is the topic of discussion in Chapter 6. Quantization noise and design of LP, HP, BP, BR digital filters using MATLAB is done in this chapter.

Chapter 7 gives a detailed idea of modern communication systems. Multiple Input Multiple Output (MIMO) Systems and the tricks and techniques in the design of MIMO Systems are studied. Signal Processing in Wireless Communication is the subject matter of Chapter 8. CCI, ACI, ISI, as well as 5G and 6G mobile communication concepts are discussed.

Principles of advanced communication systems like SDMA, TDMA, FDMA, CDMA, OFDM, Software Defined Radio, Cognitive Radio Networks, and so on are discussed in Chapter 9. To some extent, satellite communication is also discussed. At the end modern wireless communication technologies such as IoT, RFID, LoRaWAN, WiFi HaLow, CPS, and Big Data are also studied.

In Chapter 10, a very relevant and recent topic is discussed. Machine Learning for Communication Systems is introduced. Machine Learning and Deep Learning techniques have wide applications in communication systems design. Software-Defined Radio (SDR) Systems, Spectrum Sensing, and Automated Modulation Sensing are also form part of this chapter.

We follow a as-simple-as-possible style of introducing the fundamental concepts in the entire textbook. A large number of solved problems, short-term projects/assignments, and end-of-chapter exercises are also included throughout the textbook. The author expects that the style of presentation and the exercises and highly student-friendly. Several graded examples are given throughout the text.

The book contains a number of figures and tables, which make the content more clear and lucid. A detailed list of further reading materials is given at the end of each chapter, which can help the advanced learner.

ACKNOWLEDGMENTS

The author would like to place on record his immense thanks to his students, faculty colleagues, and the reviewers who have contributed considerably in the successful completion of this book. A special word of thanks to the Universities Press (India) Private Limited, Hyderabad, for granting permission to use under copy right; in certain portions from a previous book authored by me; in Chapter 4. This project is supported by the MathWorks® Inc., under their bookprogram. Special thanks to The MathWorks for providing all the free software licenses so promptly. Thanks are also due to the publishers (CRC Press) for taking up this project and completing it in a short time. The author would also like to appreciate the moral support rendered by his wife Sobha and daughters Anuja and Aabha during the entire period of the preparation of the manuscript.

K.C.Raveendranathan
ravi@cet.ac.in

About the author

K C Raveendranathan holds a bachelor's degree in Electronics and Communication Engineering (University of Kerala, 1984), Masters in Electrical Communication Engineering (IISc Bangalore, 1993), and Doctorate in Wireless Communication (University of Kerala, 2011). He had over 30 years of teaching experience and 5 years of industrial experience. He has over 30 publications to his credit including 5 books, book chapters, and several peer-reviewed journals published in national/international journals and conference proceedings. He started his profession as a design engineer in the Broadcast Transmitters R&D of Bharat Electronics Limited, Bangalore in 1985. He later joined as a faculty member in College of Engineering Trivandrum, in the department of Electronics and Communication Engineering in June 1988. After meritorious service as faculty member in various government engineering colleges in the state of Kerala, he retired as Principal from College of Engineering Trivandrum in March 2018. He further worked as Principal in Rajadhani Institute of Engineering and Technology Trivandrum till November 2020. He is a senior member of the IEEE, a life fellow of the IETE and IE (I), a Chartered Engineer of IE (I), senior member of CSI, and a life member of the ISTE. He is an active volunteer of the IETE, IE (I), and the IEEE.

1 Introduction to Signals, Systems, and Processing

In this chapter, we introduce the concepts of signals, systems, and their processing. The terms signal and function (mathematical) are often used in the same sense and to refer the same entities. We begin our discussion with a comprehensive introduction to signals. We also discuss various elementary signals and concepts of analog and digital signal processing. Also, the Linear-Time Invariant (LTI) systems, convolution principle, and the Non-Linear Time-Variant (NLTV) systems are also introduced. We conclude the chapter by giving pointers to the recent developments in signal processing.

1.1 SIGNALS AND SYSTEMS

A system is a physical entity that takes an input and changes it to an output. The input and the output are both signals. Signal is a set of information or data. Examples for signals are television or telephone signals, daily or monthly sales of a business firm, or the prices of stock market. Systems process data or signals and may modify them or extract additional information from them. Mathematically, a signal is a function of either time or space, which are both independent variables. When the independent variable is time, the signal is called a *temporal signal*. When the independent variable is space, the signal is called a *spatial signal*. Examples for temporal signals are speech, audio, video, and so on. A still picture (graphics) is an example for spatial signal. Depending on the number of independent variables, one can classify the signal as *one-dimensional* or *multi-dimensional*. A signal can be, by its very nature, electrical or otherwise. An acoustic signal, for example, is a one-dimensional non-electrical signal. A microphone converts the acoustic signal to an electrical signal. It may be noted that the microphone is an example for a system, as it converts one signal to another. The terms, *signal* and *function*, are often used in the same sense.

A temporal signal can be expressed as $x(t)$ or $y(t)$ where t is the independent variable, time. A graphical signal can be expressed as $p(x,y)$, which is a spatial signal, where x and y are the two axes. Here, p is the light intensity variation of the pixels, which is a function of both x and y. A uni-dimensional temporal signal is illustrated in figure 1.1.

1.1.1 ELEMENTARY SIGNALS-UNIT IMPULSE, UNIT STEP, AND UNIT RAMP

Elementary signals are the ones that can be used to form more complex signals and are interrelated. Later in this chapter, we will show how elementary signals can be used to build more complex signals. Mathematically, the unit impulse function is

DOI: 10.1201/9781003527589-1

1

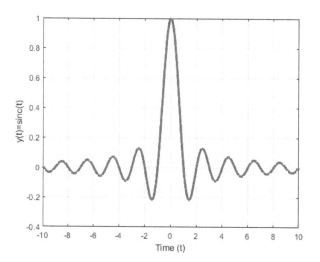

Figure 1.1   A Uni-dimensional Temporal Signal

defined as

$$\delta(t) \ = \ 0, for \ t \neq 0.$$

$$\int_{-\infty}^{\infty} \delta(t)dt \ = \ 1$$

The unit impulse function can be visualized as a tall, narrow rectangular pulse of unit area. The unit step function is defined as

$$u(t) \ = \ 0, \ t < 0$$
$$\ = \ 1, \ t > 0.$$

The unit step function has a discontinuity at $t = 0$. The unit impulse function and the unit step function are related as:

$$u(t) = \int_{\tau=-\infty}^{t} \delta(\tau)d\tau$$

Also, we have:

$$\delta(t) = \frac{d}{dt}\{u(t)\}$$

The unit ramp function is defined as:

$$r(t) \ = \ 0, \ t < 0$$
$$\ = \ t, \ t \geq 0$$

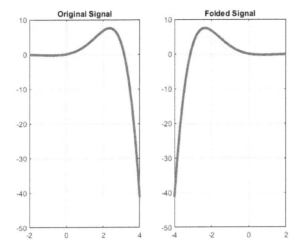

Figure 1.2 A signal and Its Folded Version.

The unit ramp and unit step functions are related as:

$$r(t) \quad = \quad \int_{\tau=-\infty}^{t} u(\tau)d\tau$$

$$u(t) \quad = \quad \frac{d}{dt}\{r(t)\}$$

1.1.2 BASIC OPERATIONS ON INDEPENDENT VARIABLE OF SIGNALS-FOLDING, SCALING, AND SHIFTING

These three operations are done on the independent variable. Signal folding, or time reversal, results when the signal is reflected wrt the independent axis $(x = 0)$. Thus, one can obtain the folded signal by replacing the independent variable x by $-x$. The original signal $f(x)$ and the folded signal $f(-x)$ are illustrated in figure 1.2. A linear scale change of the independent variable (which is time for a temporal signal) is termed as time scaling. Mathematically, time scaling is represented as:

$$y(t) = x(at), \ a \neq 0.$$

Note that time scaling results in *signal compression or down sampling* when $a > 1$; and *signal expansion or up sampling* when $a < 1$. A typical signal and its compressed version are illustrated in figure 1.3. Shifting of a signal involves changing its independent coordinates. Mathematically, shifting a signal $x(t)$ *-delaying or advancing-* by T_0 is represented as:

$$y(t) = x(t - T_0); T_0 > 0$$

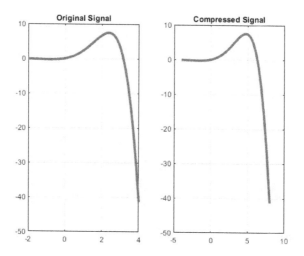

Figure 1.3   A signal and Its Compressed Version.

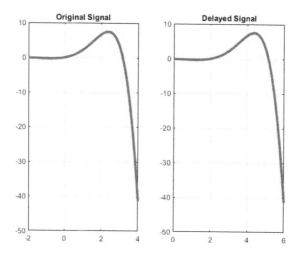

Figure 1.4   A signal and Its Delayed Version.

When $T_0 > 0$ the signal is delayed and when $T_0 < 0$, it is advanced. A signal and its delayed version is shown in figure 1.4. When it is required to process a signal with more than one operator (shifting, scaling, and folding), the following order is to be maintained to generate the correct output signal:

1. Time shift (delay or advance) the signal.
2. Time scale next.
3. Time fold last.

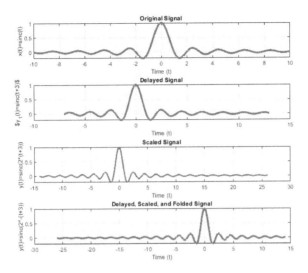

Figure 1.5 A signal and Its shifted, Scaled, and Folded Versions.

Example 1.1. A signal $x(t) = \sin(\pi * t)/(\pi * t)$, for $-10 < t < 10$ is to be transformed as $y(t) = x(-a * (t + t_0))$ where $a = 2$ and $t_0 = 3$. Obtain and plot the transformed signal.

Solution. As depicted earlier, we have to delay the signal $x(t)$ by $t_0 = 3$ first; then scale it by a factor $a = 2$, and finally fold it. Hence, the transformed signal after shifting will be

$$y_1(t) = \sin(\pi \times (t + 3))/(\pi \times (t + 3))$$

After time scaling the new signal will be

$$y_2(t) = \sin(2\pi \times (t + 3))/(2\pi \times (t + 3))$$

Finally after signal folding, the ultimate signal is

$$y(t) = \sin(2\pi \times (-(t + 3)))/(2\pi \times (-(t + 3)))$$

The signals $x(t), y_1(t), y_2(t)$, and $y(t)$ are illustrated in figure 1.5.

1.1.3 OPERATIONS ON DEPENDENT VARIABLE-LEVEL SHIFTING AND AMPLITUDE SCALING

Level Shifting and *Amplitude Scaling* are the two possible operations that may be done on the dependent variable. Mathematically, *Level Shifting* can be represented as:

$$y(t) = x(t) + a; a \neq 0$$

The *Amplitude Scaling* is mathematically expressed as:

$$y(t) = a \times x(t)$$

1.2   SIGNAL PROCESSING

Signals can be either *analog* or *digital*. Hence we have two processing methods-
analog signal processing or digital signal processing.

1.2.1   ANALOG SIGNAL PROCESSING

A signal for which the independent variable is *continuous* is termed as an *analog
signal*. For analog signals, the dependent variables take real values. Analog signals
have a continuous trail. Examples for analog signals are sinusoidal current and voltage signals. Processing of analog signals in the analog domain is referred as *analog
signal processing*. Systems that process analog signals are termed as *analog systems*.
Electronic circuits such as analog filters are examples for analog signal processing
systems. Analog filters and other analog systems have been in vogue for a long time.
We will discuss analog signal processing in greater detail later in this book.

1.2.2   DIGITAL SIGNAL PROCESSING

For digital signals, the independent variable takes only discrete values and the dependent variable takes only integer values. An analog signal can be converted to a digital
signal by a process known as *sampling*. Digital Signal Processing (DSP) has gained
wide popularity with the advent of digital computers and other digital components.
At present, the availability of inexpensive digital signal processing is also gaining
importance. Analog and digital signals are illustrated in figure 1.6. The following are
the advantages of digital signal processing compared to analog signal processing:

1. Analog systems are less accurate because of component tolerance, for example
   that of R, L, C, and active components.
2. Digital components are less sensitive to the environmental changes, noise, and
   disturbances.
3. Digital systems are most flexible as software programs can be easily modified.
   They are highly scalable.

It may be noted that, although many original communication systems were analog
(for example, analog telephones), recent technologies use digital systems and digital signal processing because of their advantages like noise immunity, possibility
of encryption (which makes the system more secure or less intrusive), bandwidth
efficiency, and the ability to use repeaters for long-distance transmission.

1.3   SYSTEMS

A system is an *entity* that processes a signal. Depending on the nature of the signal
being processed, a system can be continuous time, discrete time, or digital.

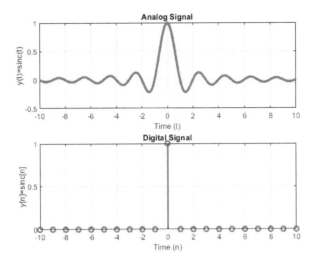

Figure 1.6 Analog and Digital Signals

1.3.1 CLASSIFICATION OF SIGNALS

There exist several classes of signals. The following are the various classes of signals within the scope of this book.

1. Continuous-time and discrete-time signals
2. Analog and digital signals
3. Periodic and aperiodic signals
4. Energy and power signals
5. Deterministic and probabilistic signals

A signal with *finite energy* is termed as an *energy signal*. A signal with finite, non-zero power is called a *power signal*. Note that power is the time average of energy. The signal $x(t) = 2\exp(-t/2)$ is an example for energy signal. The energy of a signal $x(t)$ is given by

$$E_x = \int_{-\infty}^{\infty} x^2(t)\,dt$$

and its power (if its duration is T) is

$$P_x = \frac{1}{T}\int_{-\infty}^{\infty} x^2(t)\,dt$$

Example 1.2. Classify the following signals as energy or power signals: (a) $f(t) = k\sin(t)$, *where* $0 < t < \pi$ and k is a finite, real constant. (b) $y(t) = \sum_{k=m}^{n} A_k\exp(j\omega_k t)$. Assume all frequencies to be distinct, i.e., $\omega_i \neq \omega_k$, for all $i \neq k$.

Solution.

(a) As $f(t)$ is a finite time sinusoidal signal, its energy can be calculated as:

$$E_f = \int_{t=0}^{\pi} \{f^2(t)\}dt = \int_{t=0}^{\pi} \left|k^2 \times \sin^2(t)\right| dt = k^2 \left|\frac{t}{2} - \frac{1}{4}\sin(2t)\right|_{t=0}^{\pi} = \frac{k^2\pi}{2}.$$ Hence,

the signal is an energy signal.

(b) The duration of the signal $y(t)$ is T. We have : $P_y = \frac{1}{T} \int_{t=0}^{T} \sum_{k=m}^{n} |A_k \exp(j\omega_k t)|^2 dt =$

$\sum_{k=m}^{n} |A_k|^2$. The signal is a power signal.

A signal is *periodic*, if for some positive constant T,

$$f(t) = f(t+T)$$

Note that $f(t) = f(t+T) = f(t+2T) = f(t+3T)\dots$ and the lowest value T is called the *fundamental period*.

1.3.2 CONTINUOUS TIME SYSTEMS

Continuous time systems are also known as analog systems. One can convert analog signals to discrete-time signals by the process known as sampling.

1.3.3 DISCRETE-TIME SYSTEMS

Discrete-time systems process discrete signals. In the case of discrete-time signals, the independent variable is discrete and the dependent variable is a real number. Sampling of analog signals produces discrete signals.

1.3.4 DIGITAL SYSTEMS

Digital systems process digital signals. In the case of digital signals, both the independent and dependent variables are integers. The discrete-time signals are converted to digital signals by a process called *quantization*. Digital systems gained popularity with the advent digital computers.

1.4 LINEAR TIME-INVARIANT (LTI) SYSTEMS

For a linear system, the properties of linearity (additivity) and homogeneity (scaling) are applicable. For a system to satisfy the additivity property, the following is the case: If two inputs x_1 and x_2 produce the outputs y_1 and y_2 respectively (when applied separately) then the combined input $x_1 + x_2$ must produce an output $y_1 + y_2$. A system is homogeneous if an input $x(t)$ produces an output $y(t)$, i.e., $x(t) \longrightarrow y(t)$; then an input $cx(t)$ must produce an output $cy(t)$, where c is an arbitrary real or imaginary number. Thus, by linearity we mean that if $x_1 \longrightarrow y_1$ and $x_2 \longrightarrow y_2$, then $c_1x_1 + c_2x_2 \longrightarrow c_1y_1 + c_2y_2$. A system where its parameters are non-varying with time is called *time invariant* or *constant-parameter*. For a time-invariant system, if the input is delayed by an amount T, then the output is the same as before but delayed by T

(assuming identical initial conditions). Thus, for an LTI system we have: if $x(t) \longrightarrow$
$y(t)$, then $x(t - T) \longrightarrow y(t - T)$. A system which is both *linear* and *time-invariant* is
called *linear time-invariant (LTI)*.

1.5 RESPONSE OF AN LTI CONTINUOUS TIME SYSTEM

For a linear time-invariant continuous time system, the input $x(t)$ and the output $y(t)$
related by a linear differential equation of the form

$$\frac{d^n y}{dt^n} + a_{n-1}\frac{d^{n-1}y}{dt^{n-1}} + \ldots + a_1\frac{dy}{dt} + a_0 y(t) = b_m\frac{d^m x}{dt^m} + b_{m-1}\frac{d^{m-1}y}{dt^{m-1}} + \ldots + b_1\frac{dx}{dt} + b_0 x(t)$$

where all the coefficients a_i and b_i are constants. Using the differential operational
notation D we can also write this equation as

$$(D^n + a_{n-1}D^{n-1} + \ldots + a_1 D + a_0)y(t) = (b_m D^m + b_{m-1}D^{m-1} + \ldots + b_1 D + b_0)x(t)$$

This can be shortened as $Q(D)y(t) = P(D)x(t)$ where the polynomials $Q(D)$ and
$P(D)$ are:

$$
\begin{aligned}
Q(D) &= D^n + a_{n-1}D^{n-1} + \ldots + a_1 D + a_0 \\
P(D) &= b_m D^m + b_{m-1}D^{m-1} + \ldots + b_1 D + b_0
\end{aligned}
$$

In practice, $m \leq n$, due to noise considerations. For maintaining generality, we may
assume $m = n$ in the above equation. The response of a linear system can be expressed as the sum of two components: the zero-input component and the zero-state
component. Thus: **Total response = zero-input response + zero-state response.**
The zero-input response is the system response when the input $x(t) = 0$ so that it
is the result of internal system conditions, such as energy storage and initial conditions alone. It is independent of the external input $x(t)$. In contrast, the zero-state
response is the response due to the external input $x(t)$ when the system is in *zero
state*, meaning the absence of all internal energy storages; i.e., all initial conditions
are zero.

1.5.1 THE UNIT IMPULSE RESPONSE $H(T)$

The system response to an input $x(t)$ may be found by breaking this input into narrow
rectangular pulses, and then summing up the system response to all the components.
The rectangular pulses become impulses in the limit as their widths approach zero.
Hence, the system response is the sum of its responses to various impulse components.

1.5.2 CONVOLUTION INTEGRAL

The zero-state response $y(t)$ to the input $x(t)$ is given by the convolution integral:

$$y(t) = x(t) * h(t) = \int_{\tau=-\infty}^{\infty} x(\tau)h(t - \tau)d\tau$$

Here, $h(t)$ is the response to the system to the unit impulse function.

1.5.2.1 Properties of Convolution Integral

1. The Commutative Property: Convolution operation is *commutative*; that is, $x(t) * y(t) = y(t) * x(t)$
2. The Distributive Property: As per this, $y(t) * \{x_1(t) + x_2(t)\} = y(t) * x_1(t) + y(t) * x_2(t)$
3. The Associative Property: According to this, $y(t) * \{x_1(t) * x_2(t)\} = y(t) * x_2(t) * x_1(t)$
4. The Shift Property: If $x(t) * y(t) = c(t)$, then:

$$x(t) * y(t - \tau) = c(t - \tau) \tag{1.5.1}$$
$$x(t - \tau) * y(t) = c(t - \tau) \tag{1.5.2}$$
$$x(t - \tau_1) * y(t - \tau_2) = c(t - \tau_1 - \tau_2) \tag{1.5.3}$$

5. Convolution with an Impulse: Convolution of a function with a unit impulse results in the function itself. That is, $x(t) * \delta(t) = x(t)$.
6. The Width Property: If the durations (widths) of signals $x(t)$ and $y(t)$ are T_1 and T_2, then the width of $x(t) * y(t)$ will be $T_1 + T_2$

Example 1.3. Evaluate the following integrals: (a) $\int_{-\infty}^{\infty} \delta(\tau) x(t - \tau) d\tau$ (b) $\int_{-\infty}^{\infty} \delta(\tau + 4) \exp(-\tau) d\tau$ (c) $\int_{-\infty}^{\infty} \delta(t) \exp(-j\omega t) dt$ (d) $\int_{-\infty}^{\infty} \delta(1-t)(t^3 + 2) dt$

Solution.

(a) Using the **Sampling or Sifting property** of the unit impulse function, it follows that: (a) $\int_{-\infty}^{\infty} \delta(\tau) x(t - \tau) d\tau = x(t)$ (b) $\int_{-\infty}^{\infty} \delta(\tau + 4) \exp(-\tau) d\tau = \exp(-4 \times -t) = \exp(4t)$ (c) $\int_{-\infty}^{\infty} \delta(t) \exp(-j\omega t) dt = \exp(-j\omega t)|_{t=0} = 1$ (d) $\int_{-\infty}^{\infty} \delta(1-t)(t^3 + 2) dt = (t^3 + 2)|_{t=1} = 3$

1.6 NON-LINEAR TIME VARIANT (NLTV) SYSTEMS

A system that is neither linear nor time-invariant is known as a *Non-linear Time-variant (NLTV)* system. We can not apply the convolution integral to obtain the output of such an NLTV system. Hence the study and analysis of NLTV systems are rather complex.

Nonlinear systems play a major role in the design of control systems from an engineering point of view. This is due to the fact that in practice all plants are nonlinear in nature. This is the main reason for considering the nonlinear systems in this book. Mathematically, a nonlinear system does not satisfy the superposition principle, or its output is not directly proportional to its input. The best example to explain nonlinearity is obviously a saturation. This condition exists because it is impossible to deliver an infinite amount of energy to any real-world system. Typically, the state and output equations for nonlinear systems may be written as follows:

$$\frac{dx(t)}{dt} = f\{x(t), u(t)\} \tag{1.6.4}$$

$$y(t) = g\{x(t), u(t)\} \tag{1.6.5}$$

The *Lorenz chaotic system* is an example of a nonlinear system, which can be described by the following equations:

$$\frac{dx_1(t)}{dt} = -10x_1(t) + 10x_2(t) + u(t) \tag{1.6.6}$$

$$\frac{dx_2(t)}{dt} = 28x_1(t) - x_2(t) + x_1(t)x_3(t) \tag{1.6.7}$$

$$\frac{dx_3(t)}{dt} = x_1(t)x_2(t) - \frac{8}{3}x_3(t) \tag{1.6.8}$$

Note that due to the presence of the terms $x_1(t)x_3(t)$ and $x_1(t)x_2(t)$, the above given system is non-linear.

Nonlinear systems are inherently difficult to test and to interpret the results correctly. Hence, great care must be taken to avoid preconceived ideas of how the system is likely to respond, for often this would influence the type of test being run and the processing techniques used. Time must be given for complex responses to develop and a wide range of different force, frequency, and sweep directions must be covered to ensure that all possible input regimes have been explored. Likewise, preconceived ideas about the expected type of response can lead to inappropriate conditioning of the response signals and thus mask behaviors that may lead to incorrect classification or understanding of the system. More often, ready-made simulation softwares assume a system's linearity and thus the results generated are invalid for nonlinear systems. If one has to use a commercial software suite, it is a good practice to check that all results are reproducible under different conditions such as sweep directions, level of excitation, and initial conditions.

Nonlinear systems are of great interest to engineers, physicists, and mathematicians because most real physical systems are inherently nonlinear in nature. Nonlinear equations are difficult to be solved by analytical methods and give rise to interesting phenomena such as bifurcation and chaos. Even simple nonlinear (or piecewise linear) dynamical systems can exhibit a completely unpredictable behavior, the so-called deterministic chaos. Chaos theory has been so surprising because chaos can also be found within trivial systems. To be called *chaotic*, a system should also be sensitive to initial conditions, in the sense that neighboring orbits separate exponentially fast, on average. Chaos is an aperiodic long-term behavior in a deterministic system that exhibits sensitive dependence on initial conditions.

There are a number of areas in which chaos finds its applications such as in lasers, biological systems, chemical reactors, power converters, etc.

1.7   RECENT ADVANCEMENTS IN SIGNAL PROCESSING

Signal processing is widely used in all domains of modern communication system design and implementation. Automotive Radar signal processing, signal processing associated with modern mobile communication systems (5G and 6G), and Optical signal processing are typical examples. We will consider some of these key technology enablers in this section.

Current state-of-the-art automotive radar sensors use frequency-modulated continuous wave (FMCW) radar technology, to make its hardware cost-efficient. Moreover, Multi Input Multi Output (MIMO) principle is used to increase the number of effective receiver antennas and to realize larger array apertures. The development of Advanced Driver Assistance Systems (ADAS) and Highly Automated Driving (HAD) poses several technology challenges to the researchers.

Research in information, communications and signal processing has brought about new services, applications and functions in a large number of fields which include consumer electronics, biomedical devices, and defense. These applications play an important role in advancing technologies to enhance human life in general. In terms of research in signal processing topics, there is a strong emphasis on advances in algorithmic development in the biomedical, and human-computer interfaces domain areas. More specifically, the use of deep learning for placental maturity staging is discussed as well as the use of vibration analysis for localizing impacts on surfaces for human-computer applications. In terms of communications signal processing, advances in new wireless communication such as *Non-Orthogonal Multiple Access (NOMA)* and millimeter-wave antenna design for 5G cellular mobile radio, as well as innovations in *Low-Density Parity-Check Code (LDPC)* decoding and networking coding, are currently of great research focus.

Recent advances of consumer electronics, like iPhone 5s, Google glasses, Xbox Kinect, and so on, bring people revolutionary experiences of human-machine interactions. Behind these innovations are the successes of various recent cutting-edge technologies, including voice recognition, object recognition, and motion recognition. Though many of these technologies are still far from perfect, these examples demonstrate the usefulness and importance of research on signal processing and machine learning (SPML).

SPML, however, clearly goes far beyond these well-known recognition technologies and roots deeply in many aspects of academic research and industrial development. Indeed, wherever digital sensors require signal processing, and wherever decision-making problems can be formed in a Machine Learning (ML) manner. Moreover, SPML is also deeply involved: on one hand, ML could take digital signals as raw features to learn rules; on the other hand, many statistical SP techniques are essentially ML solutions. For example, wavelet transform originally proposed for SP now has been widely used as a preprocessing for many machine learning applications, while probabilistic graphic models often used in an expert system now show their potential in image segmentation and recognition.

1.8   CONCLUDING REMARKS

We introduced the basic concepts of systems and signals in this chapter. Then various classes of signals like analog, discrete, digital, aperiodic, periodic, deterministic, and probabilistic were discussed. Various elementary signals like unit impulse, unit step, and unit ramp as well as their relations were introduced. Linear Time-invariant systems were considered and their characteristics were brought out. The convolution integral and its properties were considered. The unit impulse response of a LTI system was discussed. The zero-input and zero-state responses of an LTI continuous system were considered in detail. The processes of sampling and quantization were discussed. The concept of Non-linear Time-Variant (NLTV) system was brought out. A brief introduction to analog and digital signal processing was done. The characteristics of NLTV systems were discussed. We show why the detailed study of NLTV systems is important as there are several real-world, physical systems which are NLTV systems. Recent advancements in signal processing were also considered in the last section. A few solved exercises were also included in the chapter.

FURTHER READING

1. K.C. Raveendranathan, *Communication Systems Modelling and Simulation Using MATLAB and Simulink*, 1st Edition, Universities Press Hyderabad, 2011.
2. K.C. Raveendranathan, *Analog Communication Systems Principles and Practices*, Universities Press Hyderabad, 2014.
3. B.P. Lathi, *Signal Processing & Linear Systems*, Oxford University Press, 2008.
4. Alan V. Oppenheim, Ronald W. Schafer, and John R. Buck, *Discrete-Time Signal Processing*, 2nd Edition, Pearson, 2011.
5. Florian Engels et al., *Automotive Radar Signal Processing: Research Directions and Practical Challenges*, IEEE Journal of Selected Topics in Signal Processing, Vol. 15, No. 4, June 2021.
6. Andy W.H. Khong and Yong Liang Guan (Editors), *Recent Advances in Information, Communications and Signal Processing*, River Publishers, 2018.
7. Gelan Yang et al., *Recent Advancements in Signal Processing and Machine Learning*, Hindawi, 2014. https://doi.org/10.1155/2014/549024.

EXERCISES

1. Sketch the signals (a) $u(t-6) - u(t-10)$ (b) $u(t-4) + u(t-8)$ (c) $t^2[u(t-2) - u(t-5)]$ (d) $(t-5)[u(t-2) - u(t-5)]$
2. Sketch the following signals:
   a. $rect(\frac{t}{2})$.
   b. $\Delta(\frac{4\omega}{150})$.
   c. $rect(\frac{t-20}{8})$.
   d. $sinc(\frac{\pi\omega}{6})$.
3. Simplify the following expressions (a) $\left(\frac{\sin t}{t^2+4}\right)\delta(t)$ (b) $\left(\frac{1}{j\omega+3}\right)\delta(\omega+3)$

4. Express the following signals as the sum of their *odd* and *even* parts:
   a. $u[n]$.
   b. $\delta[n]$.
   c. $rect[n] = 1; \; for \; -4 \le n \le 4; \; = 0; \; otherwise.$
   d. $\sin(2\pi \times 100t); \; -1 \le t \le 1.$

5. Show that $\delta(t)$ is an *even function* of t. [Hint: For an even function $f(t)$, we have: $f(t) = f(-t)$]

6. Find the odd and even components of (a) $tu(t)$ (b) $\cos(\omega_0 t)u(t)$

7. For the systems described by the equations below, with the input $x(t)$ and the output $y(t)$, determine which of the systems are linear. (a) $\frac{dy}{dt} + 2y(t) = x^2(t)$ (b) $\frac{dy}{dt} + 5y(t) = t^2 x(t)$ (c) $\frac{dy}{dt} + 2y(t) = x(t)\frac{dx}{dt}$

8. For the systems described by the equations below, with the input $x(t)$ and the output $y(t)$, determine which of the systems are time-invariant. (a) $y(t) = x(t-4)$ (b) $y(t) = t^2 x(t-2)$ (c) $y(t) = \left(\frac{dx}{dt}\right)^2$ (d) $y(t) = x(-t)$

2 System Simulation Tools

In this chapter, we introduce the system simulation tools like MATLAB®, Mathematica®, GNU Octave, and Python in the context of signal processing for communication. It may be noted that MATLAB and GNU Octave have almost similar syntax. MATLAB and Mathematica are commercial simulation software packages, whereas, GNU Octave is a free software. We begin our discussion with a comprehensive introduction to MATLAB and Simulink. We also discuss various aspects of simulating communication systems using Mathematica and Octave. The various toolboxes available in MATLAB such as the Communication Systems Toolbox, Filter Design Toolbox, Image and Signal Processing Toolbox, and DSP System Toolbox are also discussed.

2.1 INTRODUCTION TO MATLAB & SIMULINK

MATLAB® is an entire software suite developed by the MathWorks, Inc, USA. MATLAB is the short form for **MAT**rix **LAB**oratory. Every year, two releases of MATLAB are made available by the MathWorks to the technical computing fraternity-the version released in March (denoted as version a) and the one released in September (denoted as version b). The MATLAB software suite includes Simulink (short for Simulation and Link), several toolboxes, blocksets and special purpose add ons for Artificial Intelligence (AI), Machine Learning, and so on. MATLAB Online enables the user community to use MATLAB online (in your desktop or mobile device) without actually installing it in the respective device.

2.1.1 AN INTRODUCTION TO MATLAB

Several standard operating systems support MATLAB; for example Microsoft Windows, Linux, or iOS. In this book, we generally use the MS Windows version of MATLAB. The opening window of MATLAB that appears, on clicking the MATLAB icon on the desktop, is illustrated in figure 2.1. MATLAB is an interpreted high level language. However, MATLAB Compiler is also available to develop standalone software for various applications.

2.1.2 BASIC FEATURES OF MATLAB

The basic data type in MATLAB is an array. Dimensioning is not necessary. Vectors of type $1 \times N$ and $N \times 1$ are the simpler data type. Two types of formats are available for integers–short or long. MATLAB is optimized for *array* or *vector* processing. Some of the features of C/C++, for example inheritance and operator overloading are carried over in MATLAB too. MATLAB provides an external interface to run Fortran and C/C++ programs within it. One can write and save MATLAB programs

DOI: 10.1201/9781003527589-2

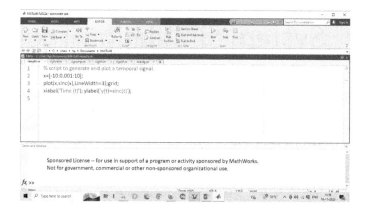

Figure 2.1   MATLAB Opening Window

in m-files, which have an extension .m, for example *sample.m*. We can use the built-in editor/debugger to develop m-files. MATLAB uses two types of m-files: *script files* and *function files*. A script file contains a sequence of MATLAB commands and is very useful when one needs to run many commands for a particular application. The script files are also called *command files*. One can execute the script file at the MATLAB command window prompt by typing its name without the extension .m. Function m-files can accept parameters whenever they are run as in C/C++ functions. A typical script file is appended below:

```
% script to plot a signal and its folded version
x=[-2:0.01:4];
y=exp(x).*sin(x);
subplot(121),plot(x,y,LineWidth=3);grid;
title('Original Signal');
nx=-fliplr(x); ny=fliplr(y);
subplot(122),plot(nx,ny,LineWidth=3);grid;
title('Folded Signal');
```

2.1.3   GETTING MATLAB HELP ONLINE

To get online help on MATLAB topics, the following commands can be used:

- help *sample*. (for on-line help on "sample.m") This command will list out all the *comments* (lines starting with a "%" character) given immediately after the *function/script definition line.* The comments that are placed elsewhere in *sample.m*; that is *not just after* the function/script definition line; will not be displayed. The programmer should make it a point to add as many comments as possible to the script/function m-file to enable readability to the user community.

Figure 2.2 Simulink Opening Menu (Simulink Start Page)

- `helpwin`. (for listing out topics).
- doc *topic*. (for html documentation.) This is the most elaborate online-help available in MATLAB.
- lookfor *topic*. (a sort of keyword search; it takes more time than the other two). The *lookfor* command will list out all instances where the string *topic* appears. It is an exhaustive search.

A detailed discussion on the MATLAB commands is available in references 1 and 2, listed at the end of this chapter.

2.1.4 INTRODUCTION TO SIMULINK

Simulink stands for System **Simu**lation and **link**. Simulink is a Graphical User Interface (GUI) based software tool. The opening menu of Simulink (Simulink Start Page) is shown in figure 2.2. It can be evoked by clicking the Simulink icon at the MATLAB Home menu (top left corner) Simulink enables users to *drag* and *drop* various building blocks from sources, sinks, user-defined functions, and other toolboxes; and simulate a complex system and study its performance. Click on the **Blank Model** icon to create a new system/subsystem model file in Simulink. Now, click on the *Library Browser*. The resulting window is shown in figure 2.3. Various model building blocks are visible now. We can choose to open a blank model from this and drag and drop desired blocks from sources and sinks and so on and save it as a new model file with extension .slx. We selected a sine wave source from the sources and a scope from the sink. Then, interconnect the source and sink, just by click on the sine source and pulling it up to the scope. The sample file is saved as "sinemodel.slx". After adjusting the parameters of the sine wave, we can run the simulation. The model and the output waveform are shown (collated) in figure 2.4.

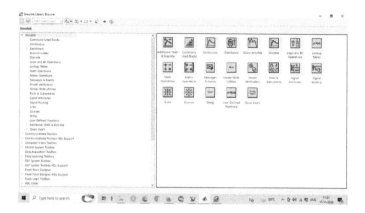

Figure 2.3   Simulink Library Browser

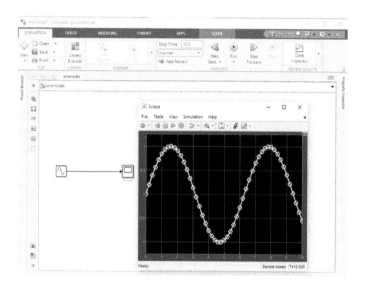

Figure 2.4   A Typical Simulink Model & Its Output Waveform

2.2   TOOLBOXES AND BLOCKSETS IN MATLAB

In MATLAB, collection of special functions is available as toolboxes. Communication Engineering students may find the following toolboxes most suitable for their use:

- *DSP System Toolbox*
- *Image Processing Toolbox*
- *Communications Toolbox*
- *Global Optimization Toolbox*
- *Fuzzy Logic Toolbox*
- *Wavelet Toolbox*
- *Communications Blockset*
- *Symbolic Math Toolbox*
- *Parallel Computing Toolbox*

- *Control System Toolbox*
- *Signal Processing Toolbox*
- *Optimization Toolbox*
- *Neural Network Toolbox*
- *Signal Processing Blockset*
- *Computer Vision Toolbox*
- *RF Blockset*
- *Statistics & Machine Learning Toolbox*
- *Deep Learning Toolbox*

An exhaustive list can be obtained from the URL:
`https://in.mathworks.com/products.html?s_tid=gn_ps` We will now consider some of the toolboxes in more detail.

2.2.1 COMMUNICATIONS TOOLBOX

This toolbox helps the design and simulation of the physical layer of communications systems. The Toolbox provides algorithms and apps for the analysis, design, end-to-end simulation, and verification of communications systems. Toolbox algorithms including channel coding, modulation, MIMO, and Orthogonal Frequency Division Multiplex (OFDM) enable the users to compose and simulate a physical layer model of their standard-based or custom-designed wireless communications system.

The toolbox provides a waveform generator app, constellation and eye diagrams, bit-error-rate, and other analysis tools and scopes for validating the designs. These tools enable the user to generate and analyze signals, visualize channel characteristics, and obtain performance metrics such as error vector magnitude (EVM). The toolbox includes Single Input Single Output (SISO) and MIMO statistical and spatial channel models. Channel profile options include Rayleigh, Rician, and WINNER II models. It also includes RF impairments, including RF nonlinearity and carrier offset and compensation algorithms, including carrier and symbol timing synchronizers. These algorithms enable users to realistically model link-level specifications and compensate for the effects of channel degradations.

Using Communications Toolbox with RF instruments or hardware support packages, one can connect the transmitter and receiver models to radio devices and verify designs with *over-the-air testing*. Physical layer features include waveform generation, source coding, error control coding, modulation, MIMO, space-time coding, filtering, equalization, and synchronization.

Usually, Communications Toolbox subsystems are used along with Simulink blocks to realize typical systems. We will consider a few such systems later in this book.

2.2.2 SIGNAL PROCESSING TOOLBOX

Signal Processing Toolbox provides functions and apps to manage, analyze, preprocess, and extract features from uniformly and non-uniformly sampled signals. The toolbox includes tools for filter design and analysis, resampling, smoothing, detrending, and power spectrum estimation. We can use the Signal Analyzer app for visualizing and processing signals simultaneously in time, frequency, and time-frequency domains. With the Filter Designer app the users can design and analyze Finite Impulse Response (FIR) and Infinite Impulse Response (IIR) digital filters. Both apps generate MATLAB scripts to reproduce or automate the design work.

Using toolbox functions, we can prepare signal datasets for AI model training by engineering features that reduce dimensionality and improve the quality of signals. We can access and process collections of files and large datasets using signal datastores. With the Signal Labeler app, the users can annotate signal attributes, regions, and points of interest to create labeled signal sets. The toolbox supports Graphic Processing Unit (GPU) acceleration in addition to C/C++ and CUDA® code generation for desktop prototyping and embedded system deployment.

2.2.2.1   Signal Analysis and Visualization

The Signal Analyzer app is an interactive tool for visualizing, measuring, analyzing, and comparing signals in the time domain, in the frequency domain, and in the time-frequency domain. The app provides a way to work with many signals of varying durations at the same time and in the same view. We can start the app by choosing it from the Apps tab on the MATLAB toolstrip, or by typing *signalAnalyzer* at the MATLAB command prompt.

2.2.2.2   Signal Generation and Preprocessing

Signal Processing Toolbox provides functions that to denoise, smooth, and detrend signals to prepare them for further analysis. There are functions to remove noise, outliers, and spurious content from data. There are also functions to generate synthetic signals such as pulses and chirps for simulation and algorithm testing. The functions available include those to generate *pulses, chirps, VCOs, sinc functions, periodic/aperiodic, and modulated signals.*

2.2.3   DSP SYSTEM TOOLBOX

DSP System Toolbox provides algorithms, apps, and scopes for designing, simulating, and analyzing signal processing systems in MATLAB and Simulink. We can model real-time DSP systems for communications, radar, audio, medical devices, IoT, and other applications. With DSP System Toolbox, the users can design and analyze FIR, IIR, multirate, multistage, and adaptive filters. The *Time Scope, Spectrum Analyzer, and Logic Analyzer* let the users dynamically visualize and measure streaming signals. For desktop prototyping and deployment to embedded processors, including ARM® Cortex® architectures, the toolbox supports C/C++ code generation. It also supports bit-accurate fixed-point modeling and Hardware Description Language (HDL) code generation from filters, Fast Fourier Transform (FFT), Inverse Fast Fourier Transform (IFFT), and other algorithms. All these algorithms are available as MATLAB functions, System objects, and Simulink blocks.

2.2.3.1 Filter Design and Analysis

Users can design and analyze a variety of digital FIR and IIR filters using DSP System Toolbox functions and apps. Some of these filters include advanced filters such as Nyquist filters, halfband filters, advanced equiripple filters, and quasi-linear phase IIR filters.

The design techniques compute the filter coefficients based on the specifications. The analysis techniques help users to validate the specifications of the designed filter. Analysis techniques include plotting the frequency response of the filter, finding the group delay of the filter, or determining if the filter is stable. Filter design and analysis are complementary and iterative. After you design a filter, analysis tools help the users to find out if the filter meets the required specifications. The toolbox provides design and analysis apps such as *filterBuilder* and *fvtool*. We can also transform filters from one form to another form using functions such as *firlp2hp, iirlp2bs, iirlp2bpc*. Refer the documentation on DSP System Toolbox for more details.

2.2.4 IMAGE PROCESSING TOOLBOX

The Image Processing Toolbox provides a comprehensive set of reference-standard algorithms and workflow apps for image processing, analysis, visualization, and algorithm development. We can perform image segmentation, image enhancement, noise reduction, geometric transformations, and image registration using deep learning and traditional image processing techniques. The toolbox supports processing of 2D, 3D, and arbitrarily large images. Image Processing Toolbox apps let the users automate common image processing workflows.

Viewing images is fundamental to image processing. The toolbox provides a number of image-processing apps to view and explore images and volumes. Using the *Image Viewer app*, we can view pixel information, pan and zoom, adjust contrast, and measure distances. Use the *Volume Viewer app* to explore volumes. The toolbox also provides visual tools for creating our own apps. One can interactively segment image data, compare image registration techniques, and batch-process large datasets. Visualization functions and apps let users to explore images, 3D volumes, and videos; adjust contrast; create histograms; and manipulate regions of interest (ROIs).

We can also speed up the algorithms by running them on multicore processors and GPUs. Many toolbox functions support C/C++ code generation for desktop prototyping and embedded vision system deployment.

2.2.5 CONTROL SYSTEM TOOLBOX

Control System Toolbox provides algorithms and apps for systematically analyzing, designing, and tuning linear control systems. We can specify the system as a transfer function, state-space, zero-pole-gain, or frequency-response model. Apps and functions, such as step response plot and Bode plot, let the users to analyze and visualize system behavior in the time and frequency domains.

We can tune compensator parameters using interactive techniques such as Bode loop shaping and the root locus method. The toolbox automatically tunes both SISO and MIMO compensators, including PID controllers. Compensators can include multiple tunable blocks spanning several feedback loops. One can tune gain-scheduled controllers and specify multiple tuning objectives, such as reference tracking, disturbance rejection, and stability margins. One can also validate the design by verifying rise time, overshoot, settling time, gain and phase margins, and other requirements.

2.2.5.1 Linear System Representation

Model objects can represent Single Input Single Output (SISO) systems or Multiple Input Multiple Output (MIMO) systems. We can represent both continuous-time and discrete-time linear systems, and systems with time delays. Basic model objects such as transfer functions and state-space models represent systems with fixed numeric coefficients. We can also build up more complex models of control systems by representing individual components as LTI models and connecting the components to model the control architecture.

2.2.5.2 Model Interconnection

Interconnecting models of components allows the user to construct models of control systems. We can conceptualize the control system as a block diagram containing multiple interconnected components, such as a plant and a controller connected in a feedback configuration. Using model arithmetic or interconnection commands, we can combine models of each of these components into a single model representing the entire block diagram.

2.2.5.3 Model Transformation

Control System Toolbox has commands for converting models from one representation to another, converting between continuous-time and discrete-time representations, and simplifying models by reducing their order.

2.2.5.4 Model Reduction

Working with lower-order models can simplify analysis and control design. Simpler models are also easier to understand and manipulate than high-order models. High-order models are obtained by linearizing complex Simulink models, interconnecting model elements, or other sources can contain states that do not contribute much to the dynamics of particular interest to the application. By using the *Model Reducer* app, the *Reduce Model Order* task in the Live Editor, or functions such as *balred* and *minreal*, we can reduce model order while preserving model characteristics that are important for the application.

2.2.6 DEEP LEARNING TOOLBOX

Deep Learning Toolbox provides a framework for designing and implementing deep neural networks with algorithms, pretrained models, and apps. We can use convolutional neural networks (ConvNets, CNNs) and long short-term memory (LSTM) networks to perform classification and regression on image, time-series, and text data. Users can build network architectures such as generative adversarial networks (GANs) and Siamese networks using automatic differentiation, custom training loops, and shared weights. With the *Deep Network Designer app*, we can design, analyze, and train networks graphically. The *Experiment Manager app* helps the users to manage multiple deep learning experiments, keep track of training parameters, analyze results, and compare code from different experiments.

One can exchange models with *TensorFlow* and *PyTorch* through the *ONNX* format and import models from *TensorFlow-Keras* and *Caffe*. The toolbox supports transfer learning with DarkNet-53, ResNet-50, NASNet, SqueezeNet, and many other pretrained models.

Users can speed up training on a single- or multiple-GPU workstation (with Parallel Computing Toolbox), or scale up to clusters and clouds, including NVIDIA® GPU Cloud and Amazon EC2® Graphic Processing Unit (GPU) instances (with MATLAB Parallel Server).

2.2.6.1 Deep Learning for Image Processing

Deep learning uses neural networks to learn useful representations of features directly from data. For example, we can use a pretrained neural network to identify and remove artifacts like noise from images.

2.2.6.2 Deep Learning with Time Series and Sequence Data

We can create and train networks for time series classification, regression, and forecasting tasks. We can train long short-term memory (LSTM) networks, to be used for sequence-to-one or sequence-to-label classification and regression problems. We can also train LSTM networks on text data using word embedding layers (requires Text Analytics Toolbox) or convolutional neural networks on audio data using spectrograms (requires Audio Toolbox).

2.2.6.3 Deep Learning Tuning and Visualization

We can interactively build and train networks, manage experiments, plot training progress, assess accuracy, explain predictions, tune training options, and visualize features learned by a network. The users can tune training options and improve network performance by sweeping hyperparameters or using Bayesian optimization. They can also use *Experiment Manager* to manage deep learning experiments that train networks under various initial conditions and compare the results. Users can monitor training progress using built-in plots of network accuracy and loss.

2.2.6.4 Deep Learning in Parallel and in the Cloud

We can scale up deep learning with multiple Graphic Processing Units (GPUs), locally or in the cloud and train multiple networks interactively or in batch jobs or train deep networks on multiple GPUs, clusters, and clouds, using *Parallel Computing Toolbox*. Users can scale up deep learning with multiple GPUs locally or on clusters, and train multiple networks interactively or in batch jobs.

2.2.6.5 Deep Learning with Big Data

Typically, training deep neural networks requires large amounts of data that often do not fit in memory. We do not need multiple computers to solve problems using data sets too large to fit in memory. Instead, we can divide the training data into mini-batches that contain a portion of the data set. By iterating over the mini-batches, networks can learn from large data sets without needing to load all data into memory at once. If the data is too large to fit in memory, use a datastore to work with mini-batches of data for training and inference. MATLAB provides many different types of datastore tailored for different applications.

2.2.7   STATISTICS AND MACHINE LEARNING TOOLBOX

Statistics and Machine Learning Toolbox provides functions and apps to describe, analyze, and model data. We can use descriptive statistics, visualizations, and clustering for exploratory data analysis, fit probability distributions to data, generate random numbers for Monte Carlo simulations, and perform hypothesis tests. Regression and classification algorithms let users to draw inferences from data and build predictive models either interactively, using the Classification and Regression Learner apps, or programmatically, using AutoML.

For multidimensional data analysis and feature extraction, the toolbox provides principal component analysis (PCA), regularization, dimensionality reduction, and feature selection methods that let users to identify variables with the best predictive power. The toolbox provides supervised, semi-supervised, and unsupervised machine learning algorithms, including support vector machines (SVMs), boosted decision trees, k-means, and other clustering methods. We can apply interpretability techniques such as partial dependence plots and local interpretable model-agnostic explanations (LIME), and automatically generate C/C++ code for embedded deployment. Many toolbox algorithms can be used on data sets that are too big to be stored in memory.

2.2.8   COMPUTER VISION TOOLBOX

Computer Vision Toolbox provides algorithms, functions, and apps for designing and testing computer vision, 3D vision, and video processing systems. We can perform object detection and tracking, as well as feature detection, extraction, and matching. Users can automate calibration workflows for single, stereo, and fish-eye cameras. For 3D vision, the toolbox supports visual and point cloud simultaneous localization and mapping (SLAM), stereo vision, structure from motion, and point cloud processing. Computer vision apps automate ground truth labeling and camera calibration workflows.

We can train custom object detectors using deep learning and machine learning algorithms such as YOLO v2, SSD, and ACF. For semantic and instance segmentation, we can use deep learning algorithms such as U-Net and Mask R-CNN. The toolbox provides object detection and segmentation algorithms for analyzing images that are too large to fit into memory. Pretrained models let users to detect faces, pedestrians, and other common objects. We can accelerate the algorithms by running them on multi-core processors and GPUs. Toolbox algorithms support C/C++ code generation for integrating with existing code, desktop prototyping, and embedded vision system deployment.

2.2.9   PARALLEL COMPUTING TOOLBOX

Parallel Computing Toolbox lets users to solve computationally and data-intensive problems using multi-core processors, Graphic Processing Units (GPUs), and computer clusters. High-level constructs–parallel for-loops, special array types, and parallelized numerical algorithms–enable users to parallelize MATLAB applications without Compute Unified Device Architecture (CUDA) or Message Passing Interface (MPI) programming. The toolbox lets users to use parallel-enabled functions in MATLAB and other toolboxes. We can use the toolbox with Simulink to run multiple simulations of a model in parallel. Programs and models can run in both interactive and batch modes.

The toolbox lets users to use the full processing power of multicore desktops by executing applications on workers (MATLAB computational engines) that run locally. Without changing the code, we can run the same applications on clusters or clouds (using MATLAB Parallel Server). We can also use the toolbox with MATLAB Parallel Server to execute matrix calculations that are too large to fit into the memory of a single machine.

2.3   FILTER DESIGN USING MATLAB

In this section, we will examine some of the functions in DSP System Toolbox that can be used for digital filter design and analysis.

2.3.1   FILTERBUILDER APP

The *filterBuilder app* design filters starting with frequency and magnitude specifications. We can open the *filterBuilder app* by typing *filterBuilder* at the MATLAB command prompt. Alternatively, we can click on the *Filter Builder app* under the **Signal Processing & Communications** window, to evoke the filter builder app. Then the Response Selection dialog box appears, listing all possible filter responses available in DSP System Toolbox. Select the Lowpass response. Now, we can start the design of the Specifications Object, and the Lowpass Design dialog box appears. This dialog box contains a Main pane, a Data Types pane, and a Code Generation pane. The specifications of the filter are generally set in the Main pane of the dialog box. The response selection dialog box is shown in figure 2.5. Now, we can view the response

Figure 2.5   The Response Selection Dialog Box Using DSP System Toolbox

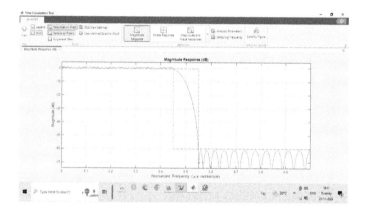

Figure 2.6   The Response of the Lowpass Filter

of the designed Lowpass filter by clicking the *View Filter Response* tab. The response of the Lowpass filter is displayed using the *Filter Visualization tool*. This is shown in figure 2.6.

The lowpass filter design is now available as a filter object *Hlp*, by clicking on the *Code Generation tab*. We can choose to generate a MATLAB function from it. The MATLAB function is listed below:

```
function Hd = getFilter
%GETFILTER Returns a discrete-time filter System object.

% MATLAB Code
```

```
% Generated by MATLAB(R) 9.12 and DSP System Toolbox 9.14.
% Generated on: 21-Nov-2023 18:54:19

Fpass = 0.45;   % Passband Frequency
Fstop = 0.55;   % Stopband Frequency
Apass = 1;      % Passband Ripple (dB)
Astop = 60;     % Stopband Attenuation (dB)

h = fdesign.lowpass('fp,fst,ap,ast', Fpass, Fstop, Apass, Astop);

Hd = design(h, 'equiripple', ...
    'FilterStructure', 'dfsymfir', ...
    'MinOrder', 'any', ...
    'StopbandShape', 'flat', ...
    'SystemObject', true);
```

The filter object Hlp can be used to lowpass filter an arbitrary input signal x to generate the filtered output y as: y=filter(Hlp,x);

2.4 SYSTEM SIMULATION USING SIMULINK

In this section, we will consider various aspects of system simulation using Simulink. We will begin with the building blocks of Simulink, its subsystems. We will then discuss typical examples (case studies) using Simulink.

2.4.1 SIMULINK BUILDING BLOCKS

Simulink can be evoked by either typing simulink at MATLAB command prompt or clicking the SIMULINK icon at the MATLAB Toolstrip Home window. Now the Simulink opening window will appear, with the following options: *Blank Model, Blank Subsystem, Blank Library, Blank Project, Folder to Project, Project from Git, Project from SVN,* and *Code Generation.* By clicking on the *Blank Model* icon, an untitled new blank model will be opened. Clicking on the **Library Browser** in the Simulation Toolstrip, the Simulink building blocks will be displayed.

2.4.2 SIMULINK SUBSYSTEMS

Apart from the generic subsystems in Simulink Sources, Sinks, and so on, a number of special subsystems are also available, which are part of the respective toolboxes. They include Communications Toolbox, Computer Vision Toolbox, Control System Toolbox, Data Acquisition Toolbox, DSP System Toolbox, Deep Learning Toolbox, Statistics and Machine Learning Toolbox, Simulink Coder, and several others. Blocks available under Communications Toolbox are shown in figure 2.7.

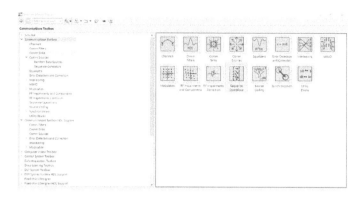

Figure 2.7   Simulink Blocks in Communications Toolbox

2.4.3 SIMULATION EXAMPLES USING SIMULINK

We will try to simulate a sinusoidal oscillator with the following transfer function:

$$T(s) = \frac{1}{s^2 + 2} \tag{2.4.1}$$

To begin with, we open a new model and select from the sources a step function and add it to the model by dragging and dropping. The parameters of the step function are set by clicking on it and editing the parameters displayed under the `Main` tab:

`step time = 0; Initial value = 0; Final value = 5; Sample time = 0;`

This is illustrated in figure 2.8. Also, check the two boxes that appear at the bottom. The output of the step function is connected to a transfer function. This is selected from `Continuous` and dragging and dropping the `Transfer Fcn` block. As before, the parameters of Transfer Function are set as shown in figure 2.9. Finally, from the `Sinks` block drag and drop a `Scope` and interconnect all the three blocks. Now run the simulation and observe the output. The output is displayed in figure 2.10.

2.5 MATLAB ONLINE

MATLAB Online provides access to MATLAB and Simulink from any standard web browser wherever there is Internet access. MATLAB Online offers cloud storage. To access MATLAB Online go to the URL: `https://in.mathworks.com/products/matlab-online.html` and login using a valid MATLAB account.

MATLAB Online cannot interact with some hardware, including instrument control. Hardware that can be accessed include:

Figure 2.8   Block Parameters of Step Function

- MATLAB Online can interact with USB webcams only through Google Chrome.
- MATLAB Online can communicate with Raspberry Pi hardware.
- MATLAB Online can interact with audio playback devices through Google Chrome.

The following are not supported by MATLAB Online:

- `Serialport()` not supported in MATLAB Online.
- Packaging tools for add-ons and MATLAB Compiler and MATLAB Compiler SDK are not supported.
- Windows-specific components like COM are not supported. `xlsread` and `xlswrite` will work in basic mode.
- Using the MEX command to build C/C++ or Fortran MEX-files is not supported.
- Files larger than 256 MB cannot be uploaded on MATLAB Online, but can be done through MATLAB Drive.
- The graphical interface to the profiler is not supported.
- Use of the shell escape bang (!) command is not fully supported.
- App Designer is only available in Google Chrome and Microsoft Edge.

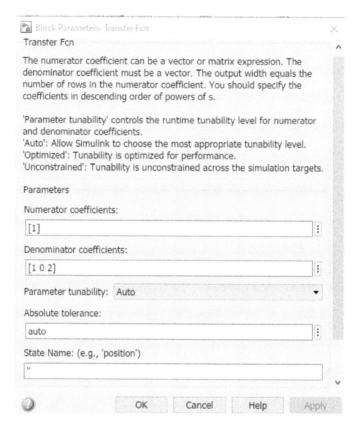

Figure 2.9   Block Parameters of Transfer Function, $T(s) = \frac{1}{s^2+2}$

Most Simulink features, including editing and simulating models, are supported.

- Simulink Online can communicate with Raspberry Pi hardware - external mode is not supported.
- Simulink Online can communicate with Parrot Minidrone hardware - deployment is not supported.
- Simulink Debugger is not supported.

The MATLAB Online opening window is shown in figure 2.11.

2.6   INTRODUCTION TO MATHEMATICA

Mathematica® is renamed as `Wolfram Language` by its originator Stephen Wolfram. The online version of the Wolfram Language is `Wolfram Alpha`. Mathematica is an extremely innovative and efficient solution for those who are looking for

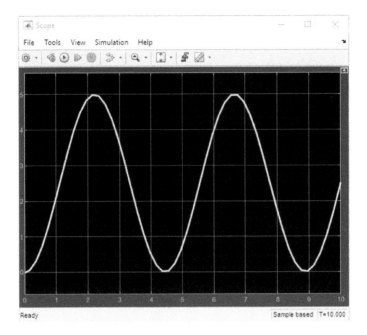

Figure 2.10   Output of Simulation: Sinusoidal Oscillator

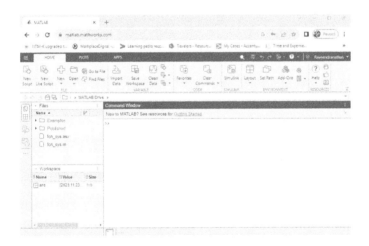

Figure 2.11   MATLAB Online Opening Window

some of the best technical computing software in the market. Its highly integrated and intuitive design allows the users to enjoy more than three decades of continuous development. This package can be used for image processing, data science, virtual visualizations, and many other unique tasks.

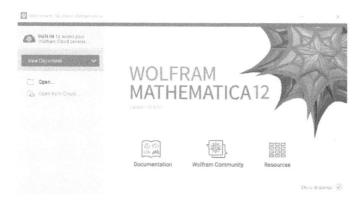

Figure 2.12 Mathematica Opening Window

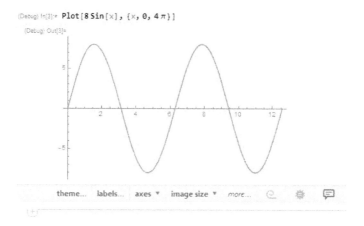

Figure 2.13 A One line Mathematica Program Code and Its Output

By clicking on the Mathematica icon, one can evoke Mathematica. The opening window is illustrated in figure 2.12. For more details on programming in Mathematica, see the references on Wolfram Language.

In the opening window, click on the **New Document** tab and select *Notebook* from the drop-down menu. Now the Notebook is open to enter Mathematica commands.

We will now write a simple program to plot $8\sin(t)$, $for\ 0 < t < 4\pi$ in Mathematica.

```
Plot[8Sin[x],{x,0,4Pi}]; (*This Code Plots 8Sin(x), for 0<x<4pi.*)
```

The resulting plot is illustrated in figure 2.13. A similar result can be obtained using the following code, too:

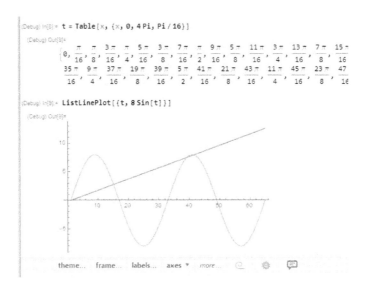

Figure 2.14 A Two line Mathematica Program Code and Its Output

```
t=Table[x,{x,0,4Pi,Pi/16}]
(*Creates a Vector t; 0<t<4Pi, in steps of pi/16. *)
ListLinePlot[{t,8Sin[t]}]; (*Plots the 8Sin(t). *)
```

The resulting plot is illustrated in figure 2.14.

2.6.1 WOLFRAM ALPHA

Wolfram Alpha is the online version of the Wolfram Language. It can be accessed through any standard web browser. The respective URL is: https://www.wolframalpha.com/

The opening window is shown in figure 2.15. Wolfram Alpha is similar in use to MATLAB Online.

2.7 INTRODUCTION TO GNU OCTAVE

GNU Octave is a free software, similar in functionality to MATLAB. GNU Octave is a high-level language primarily intended for numerical computations. It is typically used for such problems as solving linear and nonlinear equations, numerical linear algebra, statistical analysis, and for performing other numerical experiments. It may also be used as a batch-oriented language for automated data processing.

To evoke Octave, click on the GNU Octave GUI icon. The opening window is shown in figure 2.16. The syntax of GNU Octave is quite similar to that of MATLAB.

Figure 2.15 The Opening Window of Wolfram Alpha

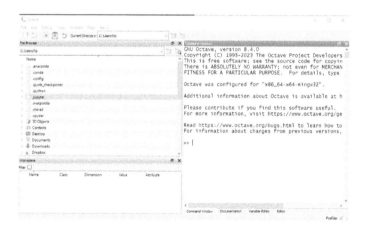

Figure 2.16 The Opening Window of GNU Octave

A simple Octave code to plot the `fft` of the *identity matrix* of order 29 is appended below:

```
plot(fft(eye(29))); axis(''square'');grid;
```

The resulting plot is shown in figure 2.17. Another GNU Octave script file to generate and plot an analog and digital signal is given below:

```
% script to generate and plot analog and digital signals.
x=[-10:0.1:10];
```

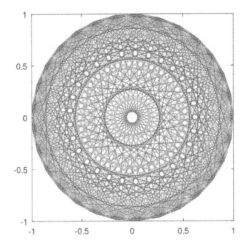

Figure 2.17   A Sample Plot Generated Using GNU Octave Code

```
subplot(211),plot(x,sinc(x));grid;
xlabel('Time (t)'); ylabel('y(t)=sinc(t)');
title('Analog Signal');
y=[-10:1:10];
subplot(212),stem(y,sinc(y));grid;
xlabel('Time (n)'); ylabel('y[n]=sinc[n]');
title('Digital Signal');
```

The resulting output is shown in figure 2.18. An exhaustive documentation on GNU Octave is available along with the software at the Help link on the opening window.

2.8   INTRODUCTION TO PYTHON

Python is the most widely used object-oriented programming language today, as per the IEEE Spectrum rankings for the past few years. Python is supported on all the main platforms and operating systems used today, such Microsoft Windows, macOS, Linux, and Android. Python is a multi-purpose programming language, which can be used for simulation, creating web pages, communicate with database systems, etc.

Python is an interpreted, high-level, general-purpose programming language. It was created by Guido van Rossum and first released in February 20, 1991. Python can be used to program Web Applications, Enterprise Applications, and Embedded Applications. Python is open source and free to use. It is ideally designed for rapid prototyping of complex applications. It has interfaces to many OS system calls and libraries and is extensible to C or C++. Python is maintained and available from **Python Software Foundation:** https://www.python.org

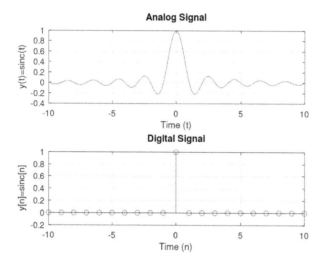

Figure 2.18   Output of GNU Octave Script File to Generate Analog and Digital Signals

Lots of Python packages exist, depending on the application. We have Python packages for Desktop GUI Development, Database Development, Web Development, Software Development, etc.

2.8.1   PYTHON LIBRARY SOFTWARES

The top 10 Python Libraries for Data Science are the following:

1. **TensorFlow** is a library for high-performance numerical computations. It is used across various scientific fields. TensorFlow is basically a framework for defining and running computations that involve tensors, which are partially defined computational objects that eventually produce a value. TensorFlow is particularly useful for the following applications:
   - Speech and image recognition
   - Text-based applications
   - Time-series analysis
   - Video detection
   See https://www.TensorFlow.org for more information.
2. **NumPy** Numerical Python is the fundamental package for numerical computation in Python; it contains a powerful N-dimensional array object. It is a general-purpose array-processing package that provides high-performance multidimensional objects called arrays and tools for working with them. NumPy also addresses the slowness problem partly by providing these multidimensional arrays as well as providing functions and operators that operate efficiently on these arrays. Applications of NumPy are:

- Extensively used in data analysis
- Creates powerful N-dimensional array
- Forms the base of other libraries, such as SciPy and SciKit-Learn
- Replacement of MATLAB when used with SciPy and Matplotlib

See `https://NumPy.org` for details.

3. **SciPy** Scientific Python is another free and open-source Python library for data science that is extensively used for high-level computations. It is extensively used for scientific and technical computations because it extends NumPy and provides many user-friendly and efficient routines for scientific calculations. Applications of SciPy include:

- Multidimensional image operations
- Solving differential equations and the Fourier transform
- Optimization algorithms
- Linear algebra

See `https://SciPy.org` for details.

4. **Pandas** Python Data Analysis is a must in the data science life cycle. It is the most popular and widely used Python library for data science, along with NumPy in Matplotlib. It is heavily used for data analysis and cleaning. Pandas provides fast, flexible data structures, such as data frame Constrained Dual Scaling (CDS), which are designed to work with structured data very easily and intuitively. Applications of Pandas are:

- General data wrangling and data cleaning
- ETL (extract, transform, load) jobs for data transformation and data storage, as it has excellent support for loading Comma Separated Variable (CSV) files into its data frame format
- Used in a variety of academic and commercial areas, including statistics, finance and neuroscience
- Time-series-specific functionality, such as date range generation, moving window, linear regression and date shifting

5. **Matplotlib** has powerful yet beautiful visualizations. It is a plotting library for Python. Because of the graphs and plots that it produces, it is extensively used for data visualization. It also provides an object-oriented API, which can be used to embed those plots into applications. See `https://matplotlib.org/` for more details.

6. **Keras** Similar to TensorFlow, Keras is another popular library that is used extensively for deep learning and neural network modules. Keras supports both the TensorFlow and Theano backends, so it is a good option if one does not want to dive into the details of TensorFlow. One of the most significant applications of Keras are the deep learning models that are available with their pretrained weights.

7. **SciKit-Learn** is a machine learning library that provides almost all the machine learning algorithms the users might need. SciKit-Learn is designed to be interpolated into NumPy and SciPy. Its applications are:

- clustering
- classification
- regression

- model selection
- dimensionality reduction

8. **PyTorch** is a Python-based scientific computing package that uses the power of graphics processing units. PyTorch is one of the most commonly preferred deep learning research platforms built to provide maximum flexibility and speed. Applications of PyTorch are:
   - PyTorch is famous for providing two of the most high-level features
   - tensor computations with strong GPU acceleration support
   - building deep neural networks on a tape-based autograd system

9. **Scrapy** is one of the most popular, fast, open-source web crawling frameworks written in Python. It is commonly used to extract the data from the web page with the help of selectors based on XPath. Applications are:
   - Scrapy helps in building crawling programs (spider bots) that can retrieve structured data from the web
   - Scrapy is also used to gather data from Application Programming Interface (APIs) and follows a `Do not Repeat Yourself` principle in the design of its interface, influencing users to write universal codes that can be reused for building and scaling large crawlers.

10. **BeautifulSoup** is another popular Python library most commonly known for web crawling and data scraping. Users can collect data that is available on some websites without a proper CSV or API, and BeautifulSoup can help them scrape it and arrange it into the required format.

2.8.2 ANACONDA

Anaconda is a distribution package, where we get Python compiler, Python packages, and the Spyder editor; all in one package. Anaconda includes Python, the Jupyter Notebook, and other commonly used packages for scientific computing and data science. Both a free version (Anaconda Distribution) and a paid version (Enterprise) Anaconda is available for Microsoft Windows, macOS, and Linux. The official website of Anaconda is: `https://www.anaconda.com`.
Spyder and the Python packages packages (NumPy, SciPy, Matplotlib, etc.) are also included in the Anaconda distribution.

2.8.3 PYTHON EDITORS

Examples for Python editors are:

- Python IDLE
- Spyder
- PyCharm
- Wing Python IDE
- Jupyter Notebook
- Visual Studio Code

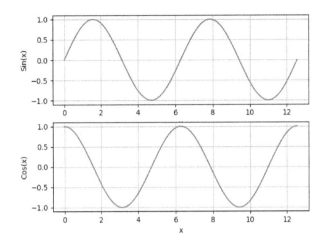

Figure 2.19   Output of Python Script to plot Sin(x) & Cos(x)

Now we will write a simple Python script file to plot sin(x) and cos(x) as two subplots on the same window. The code is appended below:

```
# A simple Python program to plot sin and cos..
import math as mt # import math library as mt
import numpy as np # import numpy library as np
import matplotlib.pyplot as plt # import matplotlib as plt
xstart=0;xstop=4*np.pi;delta=0.01
x=np.arange(xstart,xstop,delta)
y=np.sin(x);z=np.cos(x)
plt.subplot(2,1,1)
plt.plot(x,y)
plt.xlabel('x')
plt.ylabel('Sin(x)')
plt.grid()
plt.subplot(2,1,2)
plt.plot(x,z)
plt.xlabel('x')
plt.ylabel('Cos(x)')
plt.grid()
plt.show()
```

The output is shown in figure 2.19.

2.9 CONCLUDING REMARKS

In this chapter, we discussed various simulation tools which are widely used for the signal processing of communication systems. We began the discussion with an introduction to MATLAB and Simulink. After introducing the basic MATLAB features, we went on to consider the Toolboxes in MATLAB that would be most appropriate for more advanced simulations. We also introduced Simulink and its building blocks. We considered Mathematica and its features for simulating communication and signal processing. Then we considered GNU Octave, a free simulation and modeling software, that has almost similar syntax of that of MATLAB. Finally we discussed Python, which is the most popular and widely used programming tool. We introduced a few examples of simulations too in this chapter.

FURTHER READING

1. K.C. Raveendranathan, *Communication Systems Modelling and Simulation Using MATLAB and Simulink*, 1st Edition, Universities Press, Hyderabad, 2011.
2. B.P. Lathi, *Signal Processing & Linear Systems*, Oxford University Press, 2008.
3. The MathWorks Inc., *DSP System Toolbox Reference*, 2023.
4. The MathWorks Inc., *DSP System Toolbox User Guide*, 2023.
5. The MathWorks Inc., *Control System Toolbox User Guide*, 2023.
6. The MathWorks Inc., *Deep Learning Toolbox User Guide*, 2023.
7. The MathWorks Inc., *Simulink User Guide*, 2023.
8. Rafael C. Gonzalez, Richard E. Woods, & Steven L. Eddins, Digital Image Processing Using MATLAB, 3rd edition, Gatesmark Publishing, 2020.
9. Edward B. Magrab, *An Engineer's Guide to Mathematica*, John Wiley, 2014.
10. Stephen Wolfram, *An Elementary Introduction to the Wolfram Language*, 3rd Edition, 2023.
11. Hans-Petter Halvorsen, *Python Programming*, 2020.
12. Hans-Petter Halvorsen, *Python for Science and Engineering*, 2020.

EXERCISES

1. Sketch the following signals using MATLAB, Mathematica, GNU Octave and Python: (a) $u(t+6) - u(t-6)$ (b) $exp(-2t) \times [u(t-2)]$

2. Obtain the product of `eigenvalues` of the matrix, $M = \begin{bmatrix} 1 & 0 & 4 \\ 0 & 2 & -4 \\ 0 & -2 & 4 \end{bmatrix}$ using MATLAB, GNU Octave, Mathematica, and Python.

3. Use MATLAB/GNU Octave/Mathematica/Python to solve the following equations for x, y, and y as functions of the parameter a:

$$
\begin{aligned}
x - 6y + 4z &= 12a \\
6x + 3y + z &= 14a \\
5x - 2y - 6z &= 8a
\end{aligned}
$$

4. Using Simulink libraries, plot the following:
   a. $x(t) = e^t$, $\forall t \in (0,4)$.
   b. $y(t) = 3t^2 - 3t + 2$, $\forall t \in (-4, 4)$.
5. If we make $x = poly(B)$, then the MATLAB/GNU Octave command $roots(x)$ calculates the roots of the characteristic polynomial of the matrix B. Use the above two functions to obtain the *eigenvalues* of the matrix $B = \begin{bmatrix} 0 & 2 & -4 \\ -4 & 2 & 5 \\ -8 & 8 & -10 \end{bmatrix}$.
6. Plot simultaneously a sine and cosine waveform of $2kHz$ frequency and $5V$ amplitude using the scope sink tool in Simulink.
7. Using any one of the programming tools, plot a $2kHz$ sine wave and its delayed version where the delay is $\pm\pi$. Now sum the two and plot again. What is your observation?
8. Using MATLAB/GNU Octave, obtain the *quotient* and *remainder* of the function: $\frac{20x^3 + 16x^2 + 14}{4x^2 + 3x - 7}$.
9. Using the `poly` function find the roots of the polynomial $12x^3 + 152x^2 - 140x + 2560$.
10. Obtain the sum of the last 5 terms of the series $10t^4; t \in \{1,2,\ldots,10\}$.
11. Plot the periodic function $f(t) = |\sin(t)|$, $for -\pi < t < \pi$ using MATLAB/GNU Octave/Mathematica/Python.
12. Using MATLAB/Octave/Mathematica/Python generate and plot a `random` signal.
13. Using MATLAB/GNU Octave/Mathematica/Python write a function `chmon (a,b,c,d)` to calculate the *molecular weight* of a compound $C_aH_bO_cN_d$.
14. Solve the following equation for y given $x \in (1, 200)$ using MATLAB/GNU Octave/Mathematica/Python and plot.
15. Use MATLAB/GNU Octave/Mathematica/Python to obtain the coefficients of the quadratic polynomial $y = ax^2 + bx + c$ that passes through the points: $\{x,y\} = (1,4), (4,72), (5,120)$.
16. Simulate the output waveform of a `full wave bridge rectifier`, using any of the languages (MATLAB/GNU Octave/Mathematica/Python). Assume that the diode drop is 0.6V per diode.
17. Use MATLAB/Mathematica to simplify the following: $\sum\limits_{k=1}^{5} 6x^k$.
18. Compute the definite triple integral $\int\limits_0^4 \int\limits_1^4 \int\limits_1^3 xy^2z^4 dx\,dy\,dz$ using MATLAB/GNU Octave/Mathematica/Python.
19. Using MATLAB/GNU Octave/Mathematica/Python plot the *sampling function*, $s(t) = \frac{\sin t}{t}$. Also, plot $\left[\frac{\sin t}{t}\right]^2$ and $\frac{\sin t^2}{t^2}$.
20. Employing MATLAB/GNU Octave/Mathematics/Python compute the value of π based on the well-known series $\frac{\pi^2 - 8}{16} = \sum\limits_{n=0}^{\infty} \frac{1}{(2n-1)^2(2n+1)^2}$. Take only the first 100 terms of the series.

3 Signal Processing in Frequency Domain

Signal Processing or processing (transformation) of signals can be done either in frequency domain or in time domain. Signal processing in frequency domain is discussed in this chapter. It may be noted that signal processing in frequency domain has certain advantages compared to that in time domain. Frequency domain analysis is a study of signals or mathematical functions, in reference to frequency, instead of time. We will consider various *transforms* like Laplace transform, Fourier transform, z-transform, Hilbert transform, Wavelet Transform, Discrete Cosine transform (DCT), etc. that enable signal processing in frequency domain. Time domain processing of signals is not really popular due to many reasons, including the scarcity of appropriate hardware. However, recently time domain processing has also gained some attention from practicing engineers and scientists. Time domain signal processing analyzes the input signal depending on the waveforms observed over a period of time. Time domain techniques emphasize the amplitude variation in a specific time period. This has many applications in speech processing and heavy vehicle classification.

3.1 INTRODUCTION

Electronic processing of signals is often done in frequency domain due to its obvious merits. A typical example is *denoising of signals*. Noise is any unwanted, arbitrary entity that affects the performance of a system. It degrades the quality of the signals of interest and is unwarranted or undesirable. Filtering is one method for denoising. We employ a band reject filter (BRF) if we can confine the noise signal to a particular band of frequencies. Note that filters are electronic circuits that have a response that varies depending on the band of frequencies that are subjected to it. Low-pass filters, high-pass filters, band-pass filters, and band-reject filters are different classes of electronic filters. It should also be noted that the signal need not be electrical in nature for processing in frequency domain. Optical Signal Processing, for example, is done in optical frequencies using optical hardware such as lenses, prisms, gratings, and other optical devices. Optical Signal Processing is beyond the scope of this book.

3.2 TRANSFORMS USED IN FREQUENCY DOMAIN ANALYSIS

One of the fundamental principles of signal theory is that any real-time signal can be considered or expressed as a sum of fundamental periodic signals. A set of conditions, known as the *Dirichlet's Conditions* define the fundamental property for

DOI: 10.1201/9781003527589-3

expressing any arbitrary signal as a sum of sinusoidal components. The idea of employing "Trigonometric Sums" (sum of harmonically related sines and cosines) starts with Swdedish mathematician and physicist Leonhard Paul Euler (1707-1783) in 1748. Later several others like Daniel Bernoulli, Joseph Louis Lagrange, and finally Jean Baptiste Joseph Fourier developed the idea. Thus we got the condition for existence of *Fourier Series* representation and *Fourier Transform* of a signal $f(t)$, subject to the following conditions:

- $f(t)$ must have a finite number of *maxima* and *minima*.
- $f(t)$ must have a finite number of discontinuities in any given interval.
- $f(t)$ must be absolutely integrable over a period; i.e., $\int_{-\infty}^{\infty} |f(t)| dt < \infty$.

The above conditions are known as the *Dirichlet's Conditions*. As we know, signals can be either continuous time or discrete time. So we can expand the principles to obtain the transforms, which are the sinusoidal and co-sinusoidal components, for both of these types of signals. In particular, we may obtain the Laplace Transform (LT), Hilbert Transform (HT), and Fourier Transform (FT) for continuous time signals. Likewise, we can compute Discrete Fourier Transform (DFT), z-Transform (ZT), and Discrete Sine Transform (DST)/Discrete Cosine Transform (DCT) for discrete time signals. We will discuss them in greater detail later in this chapter. We will also discuss the `Inverse Transforms` of all the above in this chapter. Inverse transforms convert the transformed signals back to its original form.

3.2.1 LAPLACE TRANSFORM AND INVERSE

The Laplace transform, named after the great French mathematician and astronomer Pierre Simon de Laplace (1749-1827), breaks up a given signal $f(t)$ into exponentials e^{st} (where s is a complex variable; $s = \sigma + j\omega$). Since the signal $f(t)$ is transformed into a continuous sum of exponentials e^{st}, we need to introduce the convergence factor $e^{-\sigma t}$ in the transformation from the time domain to the frequency domain. For a signal $f(t)$, the *bilateral (two-sided) Laplace Transform (BLT)* is defined as:

$$F(s) = \int_{t=-\infty}^{\infty} f(t)e^{-st} dt \tag{3.2.1}$$

$F(s)$ is the Laplace Transform of the signal $f(t)$ and $f(t)$ is the *Inverse Laplace Transform (ILT)* of $F(s)$. It can be shown that:

$$f(t) = \frac{1}{2\pi j} \int_{c-j\infty}^{c+j\infty} F(s)e^{st} dt \tag{3.2.2}$$

where c is a constant chosen to ensure the convergence of equation 3.2.1. The above pair of equations is known as the `bilateral Laplace transform pair`. Symbolically, we can express the Laplace Transform Pair as:

$$F(s) = \mathscr{LT}[f(t)] \quad and \quad f(t) = \mathscr{LT}^{-1}[F(s)] \tag{3.2.3}$$

The Laplace Transform Pair can also be indicated as:

$$x(t) \overset{\mathscr{L}\mathscr{T}}{\longleftrightarrow} X(s)$$

Region of Convergence (ROC) of Laplace Transform is the set of values of s (the region in the complex plane) for which the integral in equation 3.2.1 converges.

Example 3.1. For a signal $f(t) = e^{-at}u(t)$, obtain the Laplace Transform and the ROC.

Solution. We have by definition, $F(s) = \int\limits_{-\infty}^{\infty} e^{-at}u(t)e^{-st}dt.$

As $u(t) = 0$ for $t < 0$ and $u(t) = 1$ for $t \geq 0$, we have:

$$F(s) = \int\limits_{0}^{\infty} e^{-at}e^{-st}dt = \int\limits_{0}^{\infty} e^{-(s+a)t}dt = -\frac{1}{s+a}e^{-(s+a)t}\Big|_{0}^{\infty} \qquad (3.2.4)$$

Note that s being complex, as $t \to \infty$, the term $e^{-(s+a)t}$ does not necessarily vanish. Now, we know that $s = \sigma + j\omega$ and $e^{-st} = e^{-\sigma t}e^{-j\omega t}$. But $\left|e^{-j\omega t}\right| = 1$ regardless of the value of ωt. Therefore, as $t \to \infty$, $e^{-st} \to 0$ only if $\sigma > 0$. Thus we have:

$$\lim_{t\to\infty} e^{-(s+a)t} = \begin{cases} 0 & \mathfrak{Re}(s+a) > 0 \\ \infty & \mathfrak{Re}(s+a) < 0 \end{cases}$$

Using this result in Equation 3.2.4 yields

$$F(s) \quad = \quad \frac{1}{s+a}, \quad \mathfrak{Re}(s+a) > 0, \ or$$

$$e^{-at}u(t) \quad \overset{\mathscr{L}\mathscr{T}}{\longleftrightarrow} \quad \frac{1}{s+a}, \quad \mathfrak{Re}\, s > -a \ (which \ is \ the \ R.O.C.)$$

3.2.2   UNILATERAL LAPLACE TRANSFORM (ULT)

The unilateral Laplace transform is a special case of the bilateral Laplace transform in which all signals are restricted to being *causal*; consequently the limits of integration for the integral in equation 3.2.1 can be taken from 0 to ∞. Hence the unilateral Laplace transform $F(s)$ of a signal $f(t)$ is defined as:

$$F(s) = \int\limits_{0^-}^{\infty} x(t)e^{-st}dt \qquad (3.2.5)$$

We choose 0^- (rather than 0^+) as the lower limit of integration to ensure the inclusion of the impulse function at $t = 0$, but also allows us to use initial conditions at 0^- (rather than at 0^+) in the solution of differential equations via the Laplace transform.

The unilateral Laplace transform simplifies the system analysis problem considerably because of its *uniqueness property*, which says that for a given $X(s)$, there is a unique inverse transform. But this simplification is attained at the cost of a limitation: we cannot analyze noncausal systems or use noncausal inputs. However, in most practical problems, this restriction is of little consequence. In practice, the term *Laplace transform* means the *unilateral Laplace transform*. A short table of unilateral Laplace transforms is given in Table 3.1:

Table 3.1

Unilateral Laplace Transform Pairs

Signal, $f(t)$	Unilateral Laplace Transform, $F(s)$
$\delta(t)$	1
$\delta(t - t_0)$	e^{-st_0}
$u(t)$	$\frac{1}{s}$
$u(t - t_0)$	$\frac{e^{-st_0}}{s}$
$tu(t)$	$\frac{1}{s^2}$
$t^n u(t)$	$\frac{n!}{s^{n+1}}$
$e^{-at}u(t),\ a > 0$	$\frac{1}{s+a},\ a > 0$
$te^{at}u(t)$	$\frac{1}{(s-a)^2}$
$t^n e^{at}u(t)$	$\frac{n!}{(s-a)^{n+1}}$
$\cos(\omega_0 t)u(t)$	$\frac{s}{s^2+\omega_0^2}$
$\sin(\omega_0 t)u(t)$	$\frac{\omega_0}{s^2+\omega_0^2}$
$e^{-at}\cos(\omega_0 t)u(t)$	$\frac{s+a}{(s+a)^2+\omega_0^2}$
$e^{-at}\sin(\omega_0 t)u(t)$	$\frac{\omega_0}{(s+a)^2+\omega_0^2}$

3.2.3 PROPERTIES OF LAPLACE TRANSFORM

Let us consider some important properties of Laplace transforms. They are useful not only in the derivation of the Laplace transforms of complex functions, but also in the solutions of linear integro-differential equations.

3.2.3.1 Linearity

The Laplace transforms obey the principle of superposition. That is: if

$$x(t) \xleftrightarrow{\mathscr{LF}} X(s),\ and$$

$$y(t) \xleftrightarrow{\mathscr{LF}} Y(s),\ then$$

$$ax(t) + by(t) \xleftrightarrow{\mathscr{LF}} aX(s) + bY(s). \tag{3.2.6}$$

3.2.3.2   Time Shifting

The time-shifting property states that: if

$$f(t) \overset{\mathscr{L}\mathscr{T}}{\longleftrightarrow} F(s), \; then$$
$$f(t-t_0) \overset{\mathscr{L}\mathscr{T}}{\longleftrightarrow} F(s)e^{-st_0}. \tag{3.2.7}$$

This property of unilateral Laplace transform holds only for positive values of t_0 because if $t_0 < 0$, the signal $f(t-t_0)u(t-t_0)$ may not be causal.

3.2.3.3   Frequency Shifting

The frequency-shifting property states that: if

$$f(t) \overset{\mathscr{L}\mathscr{T}}{\longleftrightarrow} F(s), \; then$$
$$f(t)e^{s_0 t} \overset{\mathscr{L}\mathscr{T}}{\longleftrightarrow} F(s-s_0). \tag{3.2.8}$$

3.2.3.4   Time Scaling

The time scaling property states that: if

$$f(t) \overset{\mathscr{L}\mathscr{T}}{\longleftrightarrow} F(s), \; then$$
$$f(at), \, a \geq 0 \overset{\mathscr{L}\mathscr{T}}{\longleftrightarrow} \frac{1}{a}F\left(\frac{s}{a}\right). \tag{3.2.9}$$

3.2.3.5   Differentiation in Time

The differentiation in time property states that: if

$$f(t) \overset{\mathscr{L}\mathscr{T}}{\longleftrightarrow} FX(s), \; then$$
$$\frac{df(t)}{dt} \overset{\mathscr{L}\mathscr{T}}{\longleftrightarrow} sF(s) - f(0^-). \tag{3.2.10}$$

3.2.3.6   Integration

The time integration property states that: if

$$f(t) \overset{\mathscr{L}\mathscr{T}}{\longleftrightarrow} F(s), \; then$$
$$\int_{0^-}^{t} f(\tau)d\tau \overset{\mathscr{L}\mathscr{T}}{\longleftrightarrow} \frac{F(s)}{s}, \; and \tag{3.2.11}$$

$$\int_{-\infty}^{t} f(\tau)d\tau \overset{\mathscr{L}\mathscr{T}}{\longleftrightarrow} \frac{F(s)}{s} + \frac{\int_{-\infty}^{0^-} f(\tau)d\tau}{s}. \tag{3.2.12}$$

3.2.3.7 Convolution in Time

If

$$f_1(t) \quad \overset{\mathscr{L}\mathscr{T}}{\longleftrightarrow} \quad F_1(s), \text{ and}$$

$$f_2(t) \quad \overset{\mathscr{L}\mathscr{T}}{\longleftrightarrow} \quad F_2(s), \text{ then}$$

$$f_1(t) * f_2(t) \quad \overset{\mathscr{L}\mathscr{T}}{\longleftrightarrow} \quad F_1(s)F_2(s). \tag{3.2.13}$$

3.2.3.8 Convolution in Frequency

If

$$f_1(t) \quad \overset{\mathscr{L}\mathscr{T}}{\longleftrightarrow} \quad F_1(s), \text{ and}$$

$$f_2(t) \quad \overset{\mathscr{L}\mathscr{T}}{\longleftrightarrow} \quad F_2(s), \text{ then}$$

$$f_1(t)f_2(t) \quad \overset{\mathscr{L}\mathscr{T}}{\longleftrightarrow} \quad \frac{1}{2\pi j}[F_1(s) * F_2(s)]. \tag{3.2.14}$$

The important properties of unilateral Laplace transforms are tabulated in Table 3.2:

Table 3.2
Properties of Unilateral Laplace Transforms

Property	Signal, $f(t)$	Unilateral Laplace Transform, $F(s)$
Linearity	$fx(t) + gy(t)$	$aF(s) + bG(s)$
Time Shifting	$f(t - t_0)u(t - t_0)$	$e^{-st_0}F(s)$
Frequency Shifting	$e^{s_0 t}f(t)$	$F(s - s_0)$
Differentiation in Time	$\frac{df(t)}{dt}$	$sF(s) - x(0^-)$
	$\frac{d^2 f(t)}{dt^2}$	$s^2 F(s) - sx(0^-) - \dot{f}(0^-)$
	$\frac{d^n f(t)}{dt^n}$	$s^n F(s) - \sum_{k=1}^{n} s^{n-k} f^{k-1}(0^-)$
Differentiation in Frequency	$-tf(t)$	$\frac{dF(s)}{ds}$
Time Integration	$\int_{0^-}^{t} f(\tau)d\tau$	$\frac{1}{s}F(s)$
	$\int_{-\infty}^{t} f(\tau)d\tau$	$\frac{1}{s}F(s) + \frac{1}{s}\int_{-\infty}^{0^-} f(t)dt$
Frequency Integration	$\frac{f(t)}{t}$	$\int_{s}^{\infty} F(z)dz$
Time Scaling	$f(at), a \geq 0$	$\frac{1}{a}F\left(\frac{s}{a}\right)$
Time Convolution	$f(t) * g(t)$	$F(s)G(s)$
Frequency Convolution	$f(t)g(t)$	$\frac{1}{2\pi j}F(s) * G(s)$
Initial Value Theorem	$f(0^+)$	$\lim_{s \to \infty} sF(s)$
Final Value Theorem	$f(\infty)$	$\lim_{s \to 0} sF(s)$, [poles of $sF(s)$ in LHP]
Parseval's Theorem	$f(t)$	$\int_{t=-\infty}^{\infty} |f(t)|^2 dt = \frac{1}{2\pi j} \int_{2\pi j} |F(s)|^2 ds$

3.2.4 ANALYSIS OF LTI SYSTEMS USING LAPLACE TRANSFORMS

The major application of Laplace transforms is in the analysis and characterization of LTI systems. It follows from the convolution property of the Laplace transform. We know that the output signal, $y(t)$ of a continuous time system is the convolution of the system input signal, $x(t)$ and the impulse response of the system, $h(t)$. That is:

$$y(t) = x(t) * h(t) = h(t) * x(t).$$

Now in the Laplace transform domain, we have

$$Y(s) = X(s)H(s).$$

Moreover, if the input to an LTI system is $x(t) = e^{st}$, with s in the ROC of $H(s)$, then the output will be $H(s)e^{st}$, that is e^{st} is an *eigenfunction* of the system with *eigenvalue* equal to the Laplace transform of the impulse response.

If the ROC of $H(s)$ includes the imaginary axis, then for $s = j\omega$, $H(s)$ is the *frequency response* of the LTI system. In the broader sense the Laplace transform of the impulse response, $H(s)$ is commonly referred to as the *system function* or the *transfer function*. The output function $y(t)$ can be obtained by taking the Inverse Laplace Transform of $F(s) = H(s)X(s)$. Let us now examine some of the properties of the LTI systems in the light of the transfer function, $H(s)$.

3.2.4.1 Causality of LTI Systems

For a causal LTI system, the impulse response is zero for $t < 0$, and therefore it is right sided. Consequently, we can see that the ROC associated with a system function for a causal LTI system is a right-half plane. However, it may be noted that the converse of this statement is not necessarily true. A ROC to the right of the rightmost pole dose not ensure that a system is causal; but it guarantees only that the impulse response is right sided. However, if $H(s)$ is *rational*, then we can find out whether the system is causal simply by checking to see if its ROC is a right half plane. Hence, for an LTI system with a rational system function, *causality* of the system is equivalent to the ROC being the right half plane to the right of the right most pole.

Example 3.2. For a system with impulse response $h(t) = e^{-a|t|}$, obtain the ROC and show that it is consistent.

Solution. Since $h(t) \neq 0$, $t < 0$, the system is *not* causal. The system function is

$$H(s) = \frac{1}{s+a} - \frac{1}{s-a} = \frac{-2a}{s^2 - a^2}, \quad -a < \Re\{s\} < +a.$$

For this system, $H(s)$ is rational and has an ROC that is not to the right of the right-most pole, which is consistent with the fact that the system is not causal.

3.2.4.2 Stability of LTI System

The ROC of $H(s)$ is related to the stability of the LTI system. The Bounded Input Bounded Output (BIBO) stability of a system implies that its impulse response is *absolutely integrable*, where in the Fourier transform of the impulse response converges. As the Fourier transform of a signal is equal to the Laplace transform computed along the $j\omega$-axis, with $\sigma = 0$, we have the following observations:

- An LTI system is stable *iff* the ROC associated with its system function $H(s)$ includes the entire $j\omega$-axis; that is, $\Re e\{s\} = 0$.
- It is perfectly possible for a system to be stable or unstable and have a system function that is not rational. For instance, the system function $H(s) = \frac{e^s}{s+2}$, $\Re e\{s\} > -2$ is not rational, and its impulse response is absolutely integrable, and hence the system is stable. For a causal LTI system with a rational system function $H(s)$, the ROC is to the right of the rightmost pole. Consequently, for this system to be stable (i.e., for the ROC to include the $j\omega$-axis), the rightmost pole of $H(s)$ must be to the *left* of the $j\omega$-axis.
- A causal system with rational system function $H(s)$ is stable *iff* all of the poles of $H(s)$ lie in the left half of the s-plane–i.e., all of the poles have only negative real parts.

3.2.5 FOURIER TRANSFORM AND INVERSE

The Fourier Transform is again applicable to continuous time signals, that satisfy the Dirichlet's conditions. Fourier Transform can be considered as a special case of Laplace Transform, where the complex variable $s = j\omega$. Fourier transforms can be applicable to aperiodic signals as well, where the period is assumed to be `infinite`. Mathematically, the Fourier Transform of a signal $f(t)$, denoted as $F(j\omega)$ can be computed as:

$$F(j\omega) = \int_{-\infty}^{\infty} f(t)e^{-j\omega t}\,dt \qquad (3.2.15)$$

The signal $f(t)$ can be retrieved from the $F(j\omega)$ as:

$$f(t) = \frac{1}{2\pi}\int_{-\infty}^{\infty} F(j\omega)e^{j\omega t}\,d\omega \qquad (3.2.16)$$

A list of Fourier Transform pairs is given in Table 3.3. The properties of Fourier Transforms are listed in Table 3.4.

3.2.6 DISCRETE-TIME FOURIER TRANSFORM (DTFT) AND INVERSE

As we can express a continuous time signal, $x(t)$ as a sum of sinusoids or exponentials, we can express a discrete-time signal (a.k.a. sequence), $x[n]$ as a sum of

Table 3.3

Fourier Transform Pairs

Signal, $f(t)$	Fourier Transform, $F(j\omega)$
$\delta(t)$	1
$\delta(t-t_0)$	$e^{-j\omega t_0}$
$f(t) = 1$	$2\pi\,\delta(\omega)$
$u(t)$	$\frac{1}{j\omega} + \pi\delta(\omega)$
$tu(t)$	$\frac{1}{w^2}$
$e^{-at}u(t),\ a>0$	$\frac{1}{a+j\omega},\ a>0$
$e^{-a|t|}$	$\frac{2a}{a^2+\omega^2}$
$\sum\limits_{k=-\infty}^{\infty} a_k e^{jk\omega_0 t}$	$2\pi\sum\limits_{k=-\infty}^{\infty} a_k\delta(\omega - k\omega_0)$
$e^{j\omega_0 t}$	$2\pi\delta(\omega - \omega_0)$
$\sin(\omega_0 t)u(t)$	$\frac{\omega_0}{\omega_0^2 - \omega^2}$
$\cos(\omega_0 t)u(t)$	$\frac{j\omega}{\omega_0^2 - \omega^2}$
$\cos(\omega_0 t)$	$\pi[\delta(\omega - \omega_0) + \delta(\omega + \omega_0)]$
$\sin(\omega_0 t)$	$\frac{\pi}{j}[\delta(\omega - \omega_0) - \delta(\omega + \omega_0)]$
$\sum\limits_{n=-\infty}^{\infty} \delta(t - nT)$	$\frac{2\pi}{T}\sum\limits_{k=-\infty}^{\infty}\delta\left(\omega - \frac{2\pi k}{T}\right)$
$f(t) = \begin{cases} 1, & |t| < T_1 \\ 0, & |t| > T_1 \end{cases}$	$\frac{2\sin(\omega T_1)}{\omega} = 2T_1\,sinc\left(\frac{\omega T_1}{\pi}\right)$
$f(t) = \frac{\sin(Wt)}{\pi t} = \frac{W}{\pi}\,sinc\left(\frac{Wt}{\pi}\right)$	$F(j\omega) = \begin{cases} 1, & |\omega| < W \\ 0, & |\omega| > W \end{cases}$

discrete-time exponentials (sinusoids). The discrete-time sequence $x[n]$ is derived from $x(t)$ by sampling as mentioned earlier. An arbitrary aperiodic sequence can be considered as a limiting case of a periodic sequence with the period approaching infinity[1]. As in the case of continuous-time signals, the discrete time Fourier Transform (DTFT) and the inverse transform are defined as:

$$F(e^{j\omega}) = \sum_{n=-\infty}^{\infty} f[n]e^{-j\omega n} \tag{3.2.17}$$

$$f(t) = \frac{1}{2\pi}\int_{2\pi} F(e^{j\omega})e^{j\omega n}d\omega \tag{3.2.18}$$

Equation 3.2.17 is known as the *analysis equation* and equation 3.2.18 is known as the *synthesis equation*. Together they are known as the discrete-time Fourier transform pairs. A list of various DTFT pairs is given in Table 3.5. Now important properties of DTFTs are listed in Table 3.6.

---

[1] In general, the sampled signal (sequence) from an original signal $f(t)$ is denoted as $f[n]$.

Table 3.4

Properties of Fourier Transforms

Property	Periodic Signal	Fourier Transform
Linearity	$af(t)+bg(t)$	$aF(j\omega)+bG(j\omega)$
Time Shifting	$f(t-t_0)$	$e^{-j\omega t_0}F(j\omega)$
Time Reversal	$f(-t)$	$F(-j\omega)$
Time and Frequency Scaling	$f(\alpha t)$	$\frac{1}{|\alpha|}F\left(\frac{j\omega}{\alpha}\right)$ $\alpha\neq 0.$
Multiplication	$f(t).g(t)$	$F(j\omega)*G(j\omega)=$
		$\frac{1}{2\pi}\int\limits_{-\infty}^{\infty}F(j\theta)G(j(\omega-\theta))d\theta.$
Convolution in Time	$f(t)*g(t)$	$F(j\omega)G(j\omega)$
Differentiation in Time	$\frac{d^n}{dt}f(t)$	$(j\omega)^nF(j\omega)$
Integration	$\int\limits_{-\infty}^{t}f(t)dt$	$\frac{1}{j\omega}F(j\omega)+\pi F(0)\delta(\omega)$
Differentiation in Frequency	$tf(t)$	$j\frac{d}{d\omega}F(j\omega)$
Conjugate Symmetry	$f(t)$ real	$\begin{cases} F(j\omega) & = & F^*(-j\omega) \\ \Re\{F(j\omega)\} & = & \Re\{F(-j\omega)\} \\ \Im\{F(j\omega)\} & = & -\Im\{F(-j\omega)\} \\ |F(j\omega)| & = & |F(-j\omega)| \\ \angle F(j\omega) & = & -\angle F(-j\omega) \end{cases}$
	$f(t)$ real and even	$F(j\omega)$ real and even.
	$f(t)$ real and odd	$F(j\omega)$ purely imaginary and odd.
Parseval's Theorem	$f(t)$	$\int\limits_{-\infty}^{\infty}|f(t)|^2dt=\frac{1}{2\pi}\int\limits_{-\infty}^{\infty}|F(j\omega)|^2d\omega$

3.2.7 Z-TRANSFORM AND INVERSE

Z-transform is the discrete-time counterpart of the Laplace transform. For a discrete-time LTI system with impulse response $h[n]$, the response of the system for a complex exponential of the form z^n is given as:

$$y[n]=H(z)z^n \qquad (3.2.19)$$

where

$$H(z)=\sum_{n=-\infty}^{\infty}h[n]z^{-n} \qquad (3.2.20)$$

is the z-transform of the impulse response, $h[n]$. If we make $z=e^{j\omega}$ with ω real (in other words, $|z|=1$), the summation in equation 3.2.20 becomes the discrete-time fourier transform (DTFT) of $h[n]$. In general, the z-transform of an arbitrary sequence $f[n]$ is defined as:

$$F[z]=\sum_{n=-\infty}^{\infty}f[n]z^{-n} \qquad (3.2.21)$$

Table 3.5

Discrete-Time Fourier Transform Pairs

Sequence, $f[n]$	Discrete-Time Fourier Transform, $F(e^{j\omega})$
$\delta[n]$	1
$\delta[n-n_0]$	$e^{-j\omega n_0}$
$f[n]=1$	$2\pi \sum\limits_{k=-\infty}^{\infty} \delta(\omega - 2\pi k)$
$\sum\limits_{k=<N>} a_k e^{jk(2\pi/N)n}$	$2\pi \sum\limits_{k=-\infty}^{\infty} a_k \delta\left(\omega - \frac{2\pi k}{N}\right)$
$e^{j\omega_0 n}$	$2\pi \sum\limits_{k=-\infty}^{\infty} \delta(\omega - \omega_0 - 2\pi k)$
$\cos \omega_0 n$	$\pi \sum\limits_{k=-\infty}^{\infty} \{\delta(\omega - \omega_0 - 2\pi k) + \delta(\omega + \omega_0 - 2\pi k)\}$
$\sin \omega_0 n$	$\frac{\pi}{j} \sum\limits_{k=-\infty}^{\infty} \{\delta(\omega - \omega_0 - 2\pi k) - \delta(\omega + \omega_0 - 2\pi k)\}$
$f[n]=\begin{cases} 1, & \|n\| \le N_1, \\ 0, & N_1 < \|n\| \le N/2, \\ \text{and}: \\ f[n+N] = f[n] \end{cases}$	$2\pi \sum\limits_{k=-\infty}^{\infty} a_k \delta\left(\omega - \frac{2\pi k}{N}\right)$
$\sum\limits_{k=-\infty}^{\infty} \delta[\omega - kN]$	$\frac{2\pi}{N} \sum\limits_{k=-\infty}^{\infty} \delta\left(\omega - \frac{2\pi k}{N}\right)$
$a^n u[n]$	$\frac{1}{1-ae^{-j\omega}}$
$(n+1)a^n u[n], \|a\|<1$	$\frac{1}{(1-ae^{-j\omega})^2}$
$\frac{(n+r-1)!}{n!(r-1)!}a^n u[n], \|a\|<1$	$\frac{1}{(1-ae^{-j\omega})^r}$
$f[n]=\begin{cases} 1, & \|n\| \le N \\ 0, & \|n\| > N \end{cases}$	$\frac{\sin\{\omega(N+\frac{1}{2})\}}{\sin \omega/2}$
$f[n] = \frac{\sin Wn}{\pi n} = \frac{W}{\pi} \text{sinc}\left\{\frac{Wn}{\pi}\right\},$ $0 < W < \pi.$	$F(e^{j\omega})=\begin{cases} 1, & 0 \le \|\omega\| \le W \\ 0, & W < \|\omega\| \le \pi, \text{ and} \end{cases}$ $F(e^{j\omega})$ is periodic with period 2π.

where z is a complex variable expressed in polar form as $z = re^{j\omega}$, where r is the magnitude and ω is the angle of z. We can now rewrite equation 3.2.21 as

$$F(re^{j\omega}) = \sum_{n=-\infty}^{\infty} f[n](re^{j\omega})^{-n} = \sum_{n=-\infty}^{\infty} \{f[n]re^{-n}\} e^{-j\omega n} = \mathscr{F}\mathscr{T}\{f[n]z^{-n}\} \quad (3.2.22)$$

Thus we can see that $F(re^{j\omega})$ is the Fourier transform of the sequence $f[n]$ multiplied by a real exponential r^{-n}. Also note that for $r = 1$ or equivalently $|z| = 1$, equation 3.2.21 reduces to the Fourier transform of $f[n]$. Thus:

$$F(z)|_{z=e^{j\omega}} = F(e^{j\omega}) = \mathscr{F}\mathscr{T}\{f[n]\} \quad (3.2.23)$$

Thus the z-transform reduces to the Fourier transform when the magnitude of the transform variable z is unity, i.e., for $|z| = |e^{j\omega}| = 1$. Hence, the z-transform

Table 3.6

Properties of Discrete-Time Fourier Transforms

Property	Sequence	Discrete-Time Fourier Transform
Linearity	$af[n] + bg[n]$	$aF(e^{j\omega}) + bG(e^{j\omega})$
Time Shifting	$f[n - n_0]$	$e^{-j\omega n_0} F(e^{j\omega})$
Frequency Shifting	$e^{j\omega_0 n} f[n]$	$F(e^{j(\omega - \omega_0)})$
Time Reversal	$f[-n]$	$F(e^{-j\omega})$
Differencing in Time	$f[n] - f[n-1]$	$(1 - e^{-j\omega}) F(e^{j\omega})$
Differentiation in Frequency	$nf[n]$	$j \dfrac{dF(e^{j\omega})}{d\omega}$
Accumulation	$\displaystyle\sum_{m=-\infty}^{n} f[m]$	$\dfrac{1}{1 - e^{-j\omega}} F(e^{j\omega})$ $+ \pi F(e^{j0}) \displaystyle\sum_{k=-\infty}^{\infty} \delta(\omega - 2\pi k)$
Time Expansion	$f_{<k>}[n] = \begin{cases} f[n/k], \\ \quad if\ n = Mk \\ 0, \\ \quad if\ n \neq Mk. \end{cases}$	$F(e^{jk\omega})$
Multiplication	$f[n].g[n]$	$F(e^{j\omega}) * G(e^{j\omega})$ $= \frac{1}{2\pi} \displaystyle\int_{-\infty}^{\infty} F(e^{j\theta}) G(e^{j(\omega - \theta)}) d\theta.$
Convolution	$f[n] * g[n]$	$F(e^{j\omega}).G(e^{j\omega})$
Conjugate Symmetry	$f[n]$ real	$\begin{cases} F(e^{j\omega}) & = & F^*(e^{-j\omega}) \\ \Re\{F(e^{j\omega})\} & = \\ \quad \Re\{F(e^{-j\omega})\} \\ \Im\{F(e^{j\omega})\} & = \\ \quad -\Im\{F(e^{-j\omega})\} \\ |F(e^{j\omega})| & = \\ \quad |F(e^{-j\omega})| \\ \angle F(e^{j\omega}) & = \\ \quad -\angle F(e^{-j\omega}) \end{cases}$
	$f[n]$ real and even	$F(e^{j\omega})$ real & even.
	$f[n]$ real and odd	$F(e^{j\omega})$ purely imaginary & odd.
Parseval's Theorem	$f[n]$	$\displaystyle\sum_{n=-\infty}^{\infty} |f[n]|^2 = \frac{1}{2\pi} \int_{2\pi} |F(e^{j\omega})|^2 d\omega$

becomes the Fourier transform on the contour in the complex $z - plane$ corresponding to a circle with radius unity, and this circle is referred to as *unit circle.*

3.2.7.1   Region of Convergence (ROC)

The complex $z - plane$ is shown in figure 3.1. Note that for the convergence of *z-transform*, it is required that the Fourier transform of $f[n] r^{-n}$ converges. The range

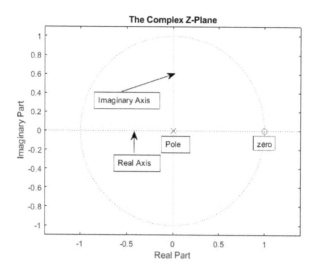

Figure 3.1   The Complex Z-Plane

of values of z over which the *z-transform* integral converges is known as the *region of convergence (ROC)*.

Example 3.3. Calculate the z-transform and the ROC of the sequence $y[n] = b^n u[n]$.

Solution. The z-transform of $y[n]$ is given by:

$$
\begin{aligned}
Y[z] &= \sum_{n=-\infty}^{\infty} y[n] z^{-n} \\
&= \sum_{n=0}^{\infty} b^n z^{-n} = \sum_{n=0}^{\infty} \left(\frac{b}{z}\right)^n \\
&= \frac{1}{1 - \frac{b}{z}} = \frac{1}{1 - bz^{-n}}; \quad |z| > |b| \\
&= \frac{z}{z - b}; \quad |z| > |b|
\end{aligned}
$$

Hence the ROC in this case, for a right-sided *causal sequence* sequence, $b^n u[n]$ is the exterior of a circle with radius b.

Example 3.4. Calculate the z-transform and the ROC of the real causal sequence $f[n] = 2 \left(\frac{1}{3}\right)^n u[n] + 4 \left(\frac{1}{5}\right)^n u[n]$.

Solution. The z-transform of $f[n]$ is given by:

Table 3.7

Z-Transform Pairs & the ROCs

Sequence, $f[n]$	Z-Transform, $F(z)$	ROC
$\delta[n]$	1	All z.
$\delta[n-n_0]$	z^{-n_0}	All z, except 0 (if $n_0 > 0$) or ∞ (if $n_0 < 0$).
$u[n]$	$\frac{z}{z-1}$	$|z| > 1$
$-u[-n-1]$	$\frac{z}{z-1}$	$|z| < 1$
$a^n u[n]$	$\frac{z}{z-a}$	$|z| > |a|$
$-a^n u[-n-1]$	$\frac{z}{z-a}$	$|z| < |a|$
$na^n u[n]$	$\frac{az}{(z-a)^2}$	$|z| > |a|$
$-na^n u[-n-1]$	$\frac{az}{(z-a)^2}$	$|z| < |a|$
$\frac{(n+1)(n+2)\ldots(2n-1)a^n}{(n-1)!}u[n]$	$\frac{z^n}{(z-a)^n}$; $z \geq 2$	$|z| > |a|$
$\frac{a^n}{n!}$	$\exp(a/z)$	$|z| > |a|$
$\cos \omega_0 n\, u[n]$	$\frac{z(z-\cos \omega_0)}{z^2-2z\cos \omega_0+1}$	$|z| > 1$
$b^n \cos \omega_0 n\, u[n]$	$\frac{z(z-\cos \omega_0)}{z^2-2bz\cos \omega_0+b^2}$	$|z| > b$
$\sin \omega_0 n\, u[n]$	$\frac{z\sin \omega_0}{z^2-2z\cos \omega_0+1}$	$|z| > 1$
$b^n \sin \omega_0 n\, u[n]$	$\frac{bz\sin \omega_0}{z^2-2bz\cos \omega_0+b^2}$	$|z| > b$

The *z-transform* of $f[n]$ is given by:

$$F_1(z) = 2\sum_{n=0}^{\infty}\left(\frac{1}{3}\right)^n z^{-n} = \frac{2}{1-\frac{1}{3}z^{-1}}, \quad |z| > \frac{1}{3}. \; and;$$

$$F_2(z) = 4\sum_{n=0}^{\infty}\left(\frac{1}{5}\right)^n z^{-n} = \frac{4}{1-\frac{1}{5}z^{-1}}, \quad |z| > \frac{1}{5}. \; therefore,$$

$$F(z) = F_1(z)+F_2(z) = \frac{2}{1-\frac{1}{3}z^{-1}} + \frac{4}{1-\frac{1}{5}z^{-1}}, \quad |z| > \frac{1}{3}.$$

The ROC is $|z| > \frac{1}{3}$ as shown above.

A list of z-transform pairs and the corresponding ROCs are given in Table 3.7.

3.2.7.2 Properties of ROC of z-Transforms

For an arbitrary sequence $f[n]$ its ROC of the *z-transform* has the following properties:

- The ROC of $F(z)$ consists of a *ring* in the $z - plane$ centered about the origin.
- The ROC does not contain any poles.

- If $f[n]$ is of finite duration, then the ROC is the entire $z - plane$, except $z = 0$ and/or $z = \infty$.
- If $f[n]$ is a right-sided sequence, and if the circle $|z| = a$ is the ROC, then all finite values of z for which $|z| > a$ will also be in the ROC.
- If $f[n]$ is a left-sided sequence, and if the circle $|z| = a$ is the ROC, then all values of z for which $0 < |z| < a$ will also be in the ROC.
- If $f[n]$ is two sided, and if the circle $|z| = a$ is in the ROC, then the ROC will consist of a ring in the $z - plane$ that includes the circle $|z| = a$.
- If the z-transform, $F(z)$ is rational, then its ROC is bounded by poles or extends to infinity.
- If the z-transform, $F(z)$ is rational, and if $f[n]$ is right sided, then its ROC is the region in the $z - plane$ outside the outermost pole-i.e., outside the circle of radius equal to the largest magnitude of poles of $F(z)$. Furthermore, if $f[n]$ is also causal, the ROC also includes $z = \infty$.
- If the z-transform, $F(z)$ is rational, and if $f[n]$ is left sided, then its ROC is the region in the $z - plane$ inside the innermost pole-i.e., inside the circle of radius equal to the smallest magnitude of poles of $F(z)$ other than any at $z = 0$ and extending inward to and possibly including $z = 0$. Particularly, if $f[n]$ is anticausal, the ROC also includes $z = 0$.

3.2.7.3 Inverse Z-Transform

The *inverse z-transform* equation is used to get back the sequence $f[n]$ for which we have calculated the *z-transform* using the equation 3.2.21. Mathematically, it is expressed as:

$$f[n] = \frac{1}{2\pi j} \oint F(z)z^{n-1}dz \qquad (3.2.24)$$

where the $\oint$ denotes integration around a counterclockwise closed contour centered at the origin and with radius r. The value of r can be chosen as any value for which $F(z)$ converges. Usually, we use the table of z-transform pairs to compute the inverse z-transform, as this formula is a bit cumbersome to calculate.

3.2.7.4 Properties of z-Transforms

In Table 3.8 we have listed the important properties of z-transforms.

3.2.7.5 Unilateral z-Transforms

The *unilateral z-transform* is particularly useful in analyzing the characteristics of causal systems specified by linear constant coefficient difference equations with nonzero initial conditions. The unilateral *z-transform* of an arbitrary sequence $f[n]$ is defined as:

$$\mathscr{F}(z) = \sum_{n=0}^{\infty} f[n]z^{-n}. \qquad (3.2.25)$$

Table 3.8
Properties of Z-Transforms

Property	Sequence, $f[n]$	z-Transform, $F(z)$
Linearity	$af[n]+bg[n]$	$aF(z)+bG(z)$
Time Shifting	$f[n-n_0]$	$z^{-n_0}F(z)$
Scaling in z-domain	$e^{j\omega_0 n}f[n]$	$F(e^{-j\omega_0}z)$
	$z_0^n f[n]$	$F\left(\frac{z}{z_0}\right)$
	$a^n f[n]$	$F(a^{-1}z)$
Time Reversal	$f[-n]$	$F(z^{-1})$
Time Expansion	$f_{<k>}[n]=\begin{cases} f[k], & n=rk, \\ 0, & n\neq rk. \end{cases}$	$F(z^k)$
Differentiation in z-domain	$nf[n]$	$-z\frac{dF(z)}{dz}$
Conjugation	$f^*[n]$	$F^*(z^*)$
Convolution	$f[n]*g[n]$	$F(z).G(z)$
First Difference	$f[n]-f[n-1]$	$(1-z^{-1})F(z)$
Accumulation	$\sum_{k=-\infty}^{n} f[k]$	$\frac{1}{1-z^{-1}}F(z)$
Initial Value	$f[0]$	$\lim_{z\to\infty} F(z)$
Final Value	$f[\infty]$	$\lim_{z\to 1}\left[\frac{z-1}{z}F(z)\right]$
Parseval's Theorem	$f[n]$	$\sum_{n=-\infty}^{\infty} |f[n]|^2 = \frac{1}{2\pi j}\oint_{2\pi j} |F(z)|^2 dz$

The unilateral *z-transform* differs from the bilateral *z-transform* in such a way that the summation is carried out only over the nonnegative values of n, irrespective of whether $f[n]$ is zero for $n<0$ or not. Thus unilateral *z-transform* of $f[n]$ can be thought of as the bilateral *z-transform* of $f[n]u[n]$. The important properties of unilateral *z-transforms* are listed in Table 3.9.

3.2.8   DISCRETE FOURIER TRANSFORM (DFT)

The numerical computation of the Fourier transform of $f(t)$ requires sample values of $f(t)$ because a digital computer can only work with *discrete data* (sequence of numbers). More than that, a computer can compute $F(j\omega)$ only at some discrete values of $j\omega$ [samples of $F(j\omega)$]. Hence, we need to relate the samples of $F(j\omega)$ to the samples of $f(t)$. This can be done using the results of *sampling theorem*.

3.2.8.1   Sampling Theorem

The sampling theorem states that a real signal $f(t)$ whose spectrum is band-limited to $B\,Hz$ ($F(j\omega)=0$ *for* $|\omega|>2\pi B$) an be reconstructed exactly (i.e., without any error) from its samples taken at a rate $\mathscr{F}_s > 2BHz$ samples per second. In other words, the minimum sampling frequency is $\mathscr{F}_s = 2B\,Hz$. This rate is called the *Nyquist Rate*.

Table 3.9

Properties of Unilateral z-Transforms

Property	Sequence, $f[n]$	z-Transform, $F(z)$
Linearity	$af[n]+bg[n]$	$aF(z)+bG(z)$
Time delay	$f[n-1]$	$z^{-1}F(z)+f[-1]$
Time advance	$f[n+1]$	$zF(z)-zf[0]$
Scaling in z-domain	$e^{j\omega_0 n}f[n]$	$F(e^{-j\omega_0}z)$
	$z_0^n f[n]$	$F\left(\frac{z}{z_0}\right)$
	$a^n f[n]$	$F(a^{-1}z)$
Time Expansion	$f_{<k>}[n] = \begin{cases} f[k], & n=rk, \\ 0, & n \neq rk. \end{cases}$	$F\left(z^k\right)$
Differentiation in z-domain	$nf[n]$	$-z\frac{dF(z)}{dz}$
Conjugation	$f^*[n]$	$F^*(z^*)$
Convolution	$f[n]*g[n]$	$F(z).G(z)$
First Difference	$f[n]-f[n-1]$	$(1-z^{-1})F(z)-f[-1]$
Accumulation	$\sum\limits_{k=-\infty}^{n} f[k]$	$\frac{1}{1-z^{-1}}F(z)$
Initial Value	$\lim\limits_{n\to 0} f[n] = f[0]$	$\lim\limits_{z\to\infty} F(z)$

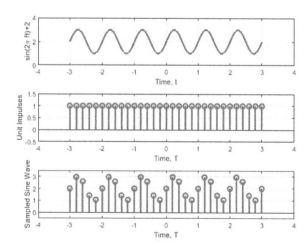

Figure 3.2 The Uniform Sampling

Sampling $f(t)$ at $\mathscr{F}_s$ can be done by multiplying $f(t)$ by an impulse train $\delta_T(t)$, as shown in figure 3.2. The MATLAB script to generate the figure is listed below:

```
% Script file to illustrate uniform sampling..
clear all; close all; clf;
```

```
n=[-3:0.01:3];
f=1;
x=sin(2*pi*f*n)+2;
subplot(311),plot(n,x,LineWidth=2);
axis([-4,4,0,4]); ylabel('sin(2\pi ft)+2');
xlabel('Time, t'); grid;
n1=[-3:0.2:3]; x1=[ones(1,length(n1))];
subplot(312),stem(n1,x1,LineWidth=2);ylabel('Unit Impulses');
xlabel('Time, T'); axis([-4,4,-0.5,1.5]); grid;
x2=sin(2*pi*f*n1)+2;
subplot(313),stem(n1,x2,LineWidth=2);ylabel('Sampled Sine Wave');
xlabel('Time, T'); axis([-4,4,-0.5,3.5]);grid;
```

Consider a continuous time signal sampled at a rate T given by $\sum_{n=-\infty}^{\infty} x[nT]\delta[t-nT]$. For brevity, we denote $x[nT]$ by $x[n]$, and with this notation, we rewrite the above expression as

$$x_n = \sum_{n=-\infty}^{\infty} x[n]\delta[t-nT] = \sum_{n=0}^{N-1} x[n]\delta[t-nT] \qquad (3.2.26)$$

The last expression in the above equation is an approximated finite sum, since in practice we have measurements for finite length of time. The values of $x(t)$ are measured at $t = 0, T, \ldots, kT, \ldots, (N-1)T$. Consequent to this, the relation for $X(j\omega)$ becomes:

$$X(j\omega) = \sum_{n=0}^{N-1} x_n e^{-jn\omega T} \qquad (3.2.27)$$

Most of the signals we use are in practice band-limited. Hence, we can use the sampling interval T as very small so that $\frac{1}{T}$ is very large. Within the band of frequency, the above equation is the Fourier transform of the signal $x(t)$. Thus, we can recover $X(j\omega)$ from the Fourier transform of the signal. However, the above equation is a finite summation, and hence it is only an approximation of $X(j\omega)$. Since the transform $X(j\omega)$ is the Fourier transform of a finite time signal of duration NT, it follows that $X(j\omega)$ is specified completely by its values at $\omega = 0, \pm\frac{2\pi}{NT}, \pm\frac{4\pi}{NT}, \ldots,$. Hence, we shall determine the values of $X(j\omega)$ only at these values of $\omega = \frac{2\pi n}{NT}$. Specifying $\omega = \frac{2\pi n}{NT}$, we can express $X(j\omega)$ as

$$X\left(\frac{j2\pi k}{NT}\right) = X[k] = \sum_{n=0}^{N-1} x[n]e^{-j\frac{2\pi nk}{N}}; \quad k = 0, 1, 2, \ldots, [N-1]. \qquad (3.2.28)$$

The function $X[k]$ is known as the Discrete Fourier Transform (DFT) of the sampled sequence $x[n]$. The equation for the Inverse Discrete Fourier Transform (IDFT) is given by

$$x[n] = \frac{1}{N}\sum_{k=0}^{N-1} X[k]e^{j\frac{2\pi nk}{N}}; \quad n = 0, 1, 2, \ldots, [N-1]. \qquad (3.2.29)$$

For notational ease, we denote $W_N = e^{-j\frac{2\pi}{N}} \Rightarrow e^{-j\frac{2\pi nk}{N}} = \left(e^{-j\frac{2\pi}{N}}\right)^{nk} = W_N^{nk}$. This symbol W is known as the *twiddle factor*. Note that both $x[n]$ and $X[k]$ are periodic. Hence, $X[k] = X[N+k]$ which simplifies the computation of DFT.

3.2.8.2   Fast Fourier Transform (FFT)

The Fast Fourier Transform is an efficient algorithm to compute the DFT. It was proposed by J.W.Cooley and J. W. Tukey, in their seminal paper, "An Algorithm for the Machine Computation of the Complex Fourier Series", published in 1965. Note that the DFT possesses the following:

1. The discrete Fourier transform is a special case of the z-transform.
2. The discrete Fourier transform can be computed efficiently using the Fast Fourier Transform algorithm.
3. The discrete Fourier transform can also be generalized to two and more dimensions.

The MATLAB/GNU Octave functions $Y = fft(x)$ and $y = ifft(X)$ implement the transform and inverse transform pair given for vectors of length N by:

$$X[k] = \sum_{n=1} Nx[n]\omega_N^{(n-1)(k-1)}$$

$$x[n] = \frac{1}{N}\sum_{k=1} NX[k]\omega_N^{-(n-1)(k-1)}$$

where $\omega_N = e^{(-2\pi j)/N}$, is an Nth root of unity.

$Y = fft(X)$ returns the discrete Fourier transform (DFT) of vector X, computed with a Fast Fourier Transform (FFT) algorithm. If X is a matrix, *fft* returns the Fourier transform of each column of the matrix. If X is a multidimensional array, *fft* operates on the first non-singleton dimension.

Example 3.5. Consider a compound sinusoidal signal containing a $60Hz$ co-sinusoid of amplitude $0.8V$ and $120Hz$ co-sinusoid of amplitude $1.2V$ and corrupted by a zero-mean random noise. Assume that the signal is sampled at 1000 Hz. Find the frequency components of the sampled signal buried in the random noise, using MATLAB/GNU Octave. Also, retrieve the signal $y(t)$ from the DFT sequences using the ifft(.) function.

Solution. The MATLAB/GNU Octave code to implement the above is listed below. Figure 3.3 illustrates the corresponding signal and the spectrum.

```
%%% MATLAB/GNU Octave code to find the frequency
%%% spectrum of a signal buried in  noise.
clear all; close all; clf;
Fs = 1000;                % Sampling frequency
T = 1/Fs;                 % Sample time
L = 10000;                % Length of signal
```

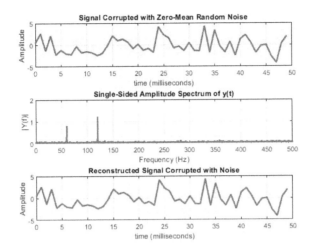

Figure 3.3 Illustration of Application of *fft(.)* Function. (a) Signal Corrupted with Zero-Mean Random Noise, $y(t)$. (b) Single-Sided Amplitude Spectrum of $y(t)$. (c) Reconstructed Signal.

```
t = [0:L-1]*T;                    % Time vector
% Sum of a 60 Hz cosinusoid and a 120 Hz cosinusoid
x = 0.8*cos(2*pi*60*t) + 1.2*cos(2*pi*120*t);
y = x + 2*randn(size(t));   % coinusoids plus Gaussian noise
subplot(311),plot(Fs*t(1:50),y(1:50),'LineWidth',2)
title('Signal Corrupted with Zero-Mean Random Noise')
xlabel('time (milliseconds)');ylabel('Amplitude');grid;
NFFT = 2^nextpow2(L); % Next power of 2 from length of y
Y = fft(y,NFFT)/L; % Computes the DFT of the signal using fft() function
f = Fs/2*linspace(0,1,NFFT/2+1);
% Plot single-sided amplitude spectrum.
subplot(312),plot(f,2*abs(Y(1:NFFT/2+1)),'LineWidth',2)
title('Single-Sided Amplitude Spectrum of y(t)')
xlabel('Frequency (Hz)'); ylabel('|Y(f)|');
axis([0 500 0 2]); grid;
y1=L*ifft(Y,NFFT);
subplot(313),plot(Fs*t(1:50),y1(1:50),'LineWidth',2);
xlabel('time (milliseconds)');ylabel('Amplitude');grid;
title('Reconstructed Signal Corrupted with Noise')
%%% end of dftcom.m
```

3.2.9   HILBERT TRANSFORM AND INVERSE

The Hilbert transform of a signal $f(t)$ is defined as the transform in which phase angle of all components of the signal is shifted by $\pm\frac{\pi}{2}$. Hilbert transform of $f(t)$ is represented by $\hat{f}(t)$ and is mathematically expressed as

$$\hat{f}(t) = \frac{1}{\pi} \int\limits_{-\infty}^{\infty} \frac{f(k)}{t-k} dk \qquad (3.2.30)$$

The Inverse Hilbert Transform of $\hat{f}(t)$ is computed as:

$$f(t) = -\frac{1}{\pi} \int\limits_{-\infty}^{\infty} \frac{\hat{f}(k)}{t-k} dk \qquad (3.2.31)$$

$f(t)$ and $\hat{f}(t)$ are known as *Hilbert Transform Pairs*. When the phase angles of all the positive frequency spectral components of a signal are shifted by $-\frac{\pi}{2}$ and the phase angles of all the negative frequency spectral components are shifted by $\frac{\pi}{2}$, then the resulting function of time is known as Hilbert transform of the given signal. In case of Hilbert transformation of a signal, the magnitude spectrum of the signal does not change, only phase spectrum of the signal is changed. Also, Hilbert transform of a signal does not change the domain of the signal.

The Hilbert transform of $f(t)$ is the convolution of $f(t)$ with the signal $\frac{1}{\pi t}$. Hence, it is the response to $f(t)$ of a linear time-invariant filter (called a Hilbert transformer) having impulse response $\frac{1}{\pi t}$. Note that the Hilbert transform of $\hat{f}(t)$ is $-f(t)$. The following are the properties of Hilbert transforms:

1. A signal and its Hilbert Transform has the same amplitude spectrum.
2. A signal and its Hilbert Transform has the same autocorrelation function.
3. A signal and its Hilbert Transform has the same energy spectral density
4. A signal and its Hilbert Transform are orthogonal to each other.
5. The Hilbert transform of the product of a low-pass signal and a high-pass signal is equal to the product of the low-pass signal and the Hilbert transform of the high-pass signal.
6. If Fourier transform exist for $f(t)$, then the Hilbert transform also exists for energy and power signals.

3.2.9.1   Applications of Hilbert Transformer

The following are the applications of Hilbert Transformers:

1. Generation of Single Side Band (SSB) modulation signals
2. Representation of band-pass signals
3. Designing of minimum phase type filters

3.2.9.2   MATLAB/GNU Octave functions to Compute Hilbert Transforms

The MATLAB/GNU Octave function hilbert(x, N) computes the $N - point$ Hilbert transform of sampled sequence x. If $N < k$, where k is the number of elements of vector x, x is truncated to a lenght N. If $N > k$, then zeros are padded to x to make its size equal to N. For more information, see the MATLAB/Octave documentation. In GNU Octave, load the package, by running **pkg load signal** command from the Octave prompt, to evoke the `hilbert(.)` function. A sample MAT-LAB/GNU Octave script to compute the Hilbert Transform function of an arbitrary signal is appended below:

```
%% MATLAB/GNU Octave Script to compute & Plot
%% the Hilbert transform function..
clear all; close all; clf;
fs = 1e4;
t = 0:1/fs:1;
x = 2.0 + cos(2*pi*120*t) + sin(2*pi*660*t) + cos(2*pi*1200*t);
y = hilbert(x);
plot(t,real(y),'-',t,imag(y),LineWidth=3);
xlim([0.01 0.03]);
legend('real','imaginary');grid;
title('Hilbert Function');
xlabel('Time (s)');
```

Figure 3.4 shows the result.

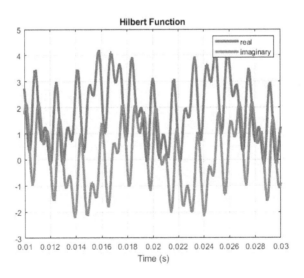

Figure 3.4 The Hilbert Transform Function

3.2.10 DISCRETE COSINE TRANSFORM (DCT) AND INVERSE

The DCT of a data sequence $X[m]$, $m = 0, 1, \ldots, (M-1)$ is defined as

$$G_x(0) = \frac{\sqrt{2}}{M} \sum_{m=0}^{M-1} X[m]$$

$$G_x(k) = \frac{2}{M} \sum_{m=0}^{M-1} X[m] \cos \frac{(2m+1)k\Pi}{2M}, \quad k = 1, 2, \ldots, (M-1). \quad (3.2.32)$$

where $G_x(k)$ is the kth DCT coefficient. It is worthwhile to note that the set of basis vectors $\left\{ \frac{1}{\sqrt{2}}, \cos \frac{(2m+1)k\Pi}{2M} \right\}$ is actually a class of Discrete Chebyshev Polynomials.

The Inverse Discrete Cosine Transform (IDCT) is defined as

$$X[m] = \frac{1}{\sqrt{2}} G_x(0) + \sum_{k=1}^{M-1} G_x(k) \cos \frac{(2m+1)k\Pi}{2M}, \quad m = 0, 1, \ldots, (M-1) \quad (3.2.33)$$

3.2.10.1 Applications of DCTs

Discrete Cosine Transforms are widely used for applications such as encoding, decoding, video, audio, multiplexing, control signals, signaling, and analog-to-digital conversion. DCTs are also commonly used for high-definition television (HDTV) encoder/decoder chips. DCT can be used in the area of image processing for the purposes of feature selection in pattern recognition; and scalar-type Wiener filtering. Its performance compares closely with that of the Karhunen-Loeve Transform (KLT), which is considered to be optimal.

3.2.10.2 Image Compression Using DCT

Image is stored or transmitted as pixel values. It can be compressed by reducing the value its every pixel contains. Image compression is basically of two types:

- Lossless compression: In this type of compression, after recovering image is exactly the same as that was before applying compression techniques and so, its quality did not get deteriorate.
- Lossy compression: In this type of compression, after recovering we can not get exactly as the original image data, and the quality of image gets significantly reduced. But this type of compression results in very high compression ratio of image data and is very useful in transmitting image over a communication network.

Discrete Cosine Transform is used in lossy image compression because it has very strong energy compaction, i.e., its large amount of information is stored in very low frequency components of a signal; and the rest in other frequency components, having very small data which can be stored by using very less number of bits (usually, at most 2 or 3 bits). To perform DCT transformation on an image, first we have to fetch image file information (pixel value in term of integer having range $0-255$) which we divide in block of 8×8 matrix and then we apply discrete cosine transform on that block of data. After applying discrete cosine transform, we will see that it is more than 90% data will be in lower frequency component.

3.2.10.3 Computation of DCT

The equation 3.2.32 can be written as

$$G_x(0) = \frac{\sqrt{2}}{M} \sum_{m=0}^{M-1} X[m]$$

$$G_x(k) = \frac{2}{M} \Re e \left\{ e^{\frac{-ik\Pi}{2M}} \sum_{m=0}^{2M-1} X[m] W^{km} \right\} \quad k = 1, 2, \ldots, (M-1). \quad (3.2.34)$$

where $W = e^{-i2\Pi/2M}$; $i = \sqrt{-1}$, and $X[m] = 0$, $m = M, M+1, \ldots, (2M-1)$. It follows that all the M DCT coefficients can be computed using a $2M$ point Fast Fourier transform (FFT). Since equations 3.2.32 and 3.2.33 are of the same form, FFT can also be used to compute the IDCT.

3.2.10.4 MATLAB/GNU Octave Functions to Compute DCT and IDCT

The function Y=dct (X, N) computes the N-point discrete cosine transform of X. If N is given, then X is padded or trimmed to length N before computing the transform. If X is a matrix, the function computes the transform along the columns of the matrix. The transform is faster if X is real-valued and has even length. Likewise, Y = idct (X, N) computes the inverse discrete cosine transform of X.

3.2.11 KARHUNEN-LOEVE TRANSFORM (KLT)

The KLT is a linear transform where the basis functions are taken from the statistics of the signal, and can thus be adaptive. It is optimal in the sense of energy compaction, i.e. it places as much energy as possible in as few coefficients as possible. The KLT is also called *Principal Component Analysis (PCA)*, is also equivalent with the *Singular Value Decomposition (SVD)*. The transform is generally not separable, and thus the full matrix multiplication must be performed. The Karhunen Loeve Transform is a key element of many signal-processing tasks, including classification and compression. The KLT of a vector x is computed as:

$$X = U^T x \quad (3.2.35)$$

and the inverse transform is computed as

$$x = UX \quad (3.2.36)$$

where the U is the basis for the transform. U is estimated from a number of x_i, $i \in \{0, 1, 2, \ldots, k\}$.

3.2.11.1 Limitations of KL Transforms

Despite its favorable theoretical properties, the KLT is not used in practice due to the following reasons:

1. Its basis functions depend on the covariance matrix of the image, and hence they have to be recomputed and transmitted for every image.
2. Perfect decorrelation is not possible, since images can rarely be modeled as realizations of ergodic fields.
3. There are no fast computational algorithms for its implementation.

3.2.12 HOUGH TRANSFORM

The Hough Transform is a technique which can be used to isolate features of a particular shape within an image. Because it requires that the desired features be specified in some parametric form, the classical Hough Transform is most commonly used for the detection of regular curves such as lines, circles, ellipses, etc. in images. A generalized Hough Transform can be employed in applications where a simple analytic description of a feature(s) is not possible. Despite its domain restrictions, the classical Hough Transform has many applications, as most manufactured parts (and many anatomical parts investigated in medical imagery) contain feature boundaries which can be described by regular curves. The main advantage of the Hough Transform technique is that it is tolerant of gaps in feature boundary descriptions and is relatively unaffected by image noise.

The Hough Transform is particularly useful for computing a global description of a feature(s) (where the number of solution classes need not be known a priori), given (possibly noisy) local measurements. The motivating idea behind the Hough Transform for line detection is that each input measurement (for example, coordinate point) indicates its contribution to a globally consistent solution (for example the physical line which gave rise to that image point). The standard Hough Transform use the parametric representation of a line, when used to detect lines in an image:

$$x\cos\theta + y\sin\theta = \rho$$

where ρ is the length of a normal from the origin to this line and θ is the orientation of ρ with respect to the X-axis. For any point (x, y) on this line, ρ and θ are constant. Note that the lines generated by the Hough Transform are infinite in length.

3.2.12.1 MATLAB/GNU Octave Functions to Compute Hough Transforms

[H, THETA, RHO] = hough(BW) computes the Standard Hough Transform (SHT) of the binary image BW. THETA (in degrees) and RHO are the arrays of ρ and θ values over which the Hough Transform matrix, H, was generated.

[H, THETA, RHO] = hough(BW,PARAM1,VAL1,PARAM2,VAL2) sets various parameters. Parameter names can be abbreviated, and are case-insensitive. While using GNU Octave, one should run the command pkg load image at Octave command prompt before evoking the hough command.

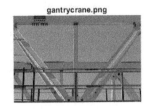

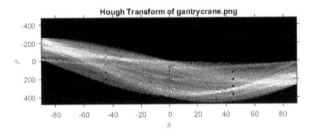

Figure 3.5   The Hough Transform of a Gray Scale Image

A sample MATLAB/GNU Octave script to get the Hough Transform of a black and white (graynscale) image is appended below. The output generated is shown in figure 3.5.

```
% Compute and display the Hough transform of the gantrycrane.png image
clear all; close all; clf;
RGB = imread('gantrycrane.png');
I = rgb2gray(RGB); % convert to intensity
BW = edge(I,'canny'); % extract edges using Canny Algorithm
[H,T,R] = hough(BW,'RhoResolution',0.5,'Theta',-90:0.5:89.5);
% display the original image
subplot(2,1,1); imshow(I);
title('gantrycrane.png');
% display the Hough Transform matrix
subplot(2,1,2);
imshow(imadjust(rescale(H)),'XData',T,'YData',R,...
   'InitialMagnification','fit');
title('Hough Transform of gantrycrane.png');
xlabel('\theta'), ylabel('\rho');
axis on, axis normal, hold on;
```

3.2.13   HANKEL TRANSFORM AND INVERSE

Hankel Transforms are integral transformations whose kernels are *Bessel Functions*. They are sometimes referred to as Bessel Transforms. While dealing with problems that show circular symmetry, Hankel Transforms may be very useful.

Laplace's partial differential equation in cylindrical coordinates can be transformed into an ordinary differential equation using the Hankel Transform. Since the Hankel Transform is the two-dimensional Fourier Transform of a circularly symmetric function, it plays an important role in Optical Data Processing.

Let $f(r)$ be a function defined for $r \geq 0$. Then the vth order Hankel Transform is defined as:

$$F_v(s) = \mathcal{H}_v\{f(r)\} = \int_0^\infty rf(r)J_v(sr)dr \qquad (3.2.37)$$

where $J_v(sr)$ is the Bessel Function of vth order. If $v > -1/2$, Hankel's repeated integral gives the inversion formula:

$$f(r) = \mathcal{H}_v^{-1}\{F_v(s)\} = \int_0^\infty sF_v(s)J_v(sr)ds \qquad (3.2.38)$$

Note that Hankel Transforms do not have as many elementary properties as do the Laplace or the Fourier Transforms. For example, because there is no simple addition formula for Bessel functions, the Hankel Transform does not satisfy any simple convolution relation.

Hankel Transforms can be used to solve a class of linear time-varying differential equations and make the solution much easier than by classical methods, especially in finding a particular solution.

3.2.14 SHORT TIME FOURIER TRANSFORM (STFT) AND INVERSE

The Short-Time Fourier Transform (STFT) is used to analyze how the frequency content of a nonstationary signal changes over time. The magnitude squared of the STFT is known as the *spectrogram time-frequency representation* of the signal. In practice, the procedure for computing STFTs is to divide a longer time signal into shorter segments of equal length and then compute the Fourier Transform separately on each shorter segment. This reveals the Fourier Spectrum on each shorter segment.

3.2.14.1 Difference between Discrete Fourier Transform and STFT

Note that the DFT has no temporal resolution (all of time is shown together in the frequency plot). In contrast, the STFT provides both temporal and frequency resolution: for a given time, we get a spectrum. This enables us to better represent signals with spectra that change over time.

3.2.14.2 Continuous Time STFT

In the continuous-time case, the function to be transformed is multiplied by a window function which is nonzero for only a short period of time. The Fourier Transform (a one-dimensional function) of the resulting signal is taken, then the window is slid along the time axis until the end resulting, in a two-dimensional representation of the signal. Mathematically, this is written as:

$$STFT\{x(t)\}(\tau,\omega) = X(\tau,\omega) = \int_{-\infty}^{\infty} x(t)w(t-\tau)e^{-j\omega t}dt \qquad (3.2.39)$$

where $w(\tau)$ is the `window function`, commonly a Hann Window or Gaussian Window centered around 0, and $x(t)$ is the signal to be transformed. $X(\tau,\omega)$ is essentially the Fourier Transform of $x(t)w(t-\tau)$, a complex function representing the phase and magnitude of the signal over time and frequency. The Inverse Continuous-Time STFT is given by:

$$x(t) = \frac{1}{2\pi} \int_{-\infty}^{\infty}\int_{-\infty}^{\infty} X(\tau,\omega)e^{j\omega t}d\tau d\omega \qquad (3.2.40)$$

In the discrete-time case, we have the following formula for the DT-STFT:

$$STFT\{x[n]\}(m,\omega) = X(m,\omega) = \sum_{n=-\infty}^{\infty} x[n]w[n-m]e^{-j\omega n} \qquad (3.2.41)$$

In this case, m is discrete and ω is continuous, but in most typical applications the STFT is performed on a computer using the Fast Fourier Transform, so both variables are discrete and quantized. The magnitude squared of the STFT yields the spectrogram representation of the power spectral density of the function:

$$spectrogram\{x(t)\}(\tau,\omega) = |X(\tau,\omega)|^2 \qquad (3.2.42)$$

In general, the window function $w(t)$ has the following properties:

1. even symmetry: $w(t) = w(-t)$.
2. non increasing for positive time: $w(t) \geq w(s)$ if $|t| \leq |s|$.
3. compact support: $w(t)$ is equal to zero when $|t|$ is large.

3.2.14.3 MATLAB Function to Compute STFT

$S = stft(X)$ returns the short-time Fourier transform (stft) of X. X can be a vector, a matrix, or a timetable. If the input has multiple channels, specify X as a matrix where each column corresponds to a channel. If X is a timetable, it must contain finite and uniformly increasing time values. For multichannel timetable input, specify X as a timetable with a single variable containing a matrix or a timetable with multiple variables, each containing a column vector. Precision can be double or single but cannot be mixed. The output S contains a two-sided and centered stft for each signal channel. S is a matrix for single-channel signals and a 3-D array for multichannel signals. Time increases across the columns and frequency increases down the rows. The third dimension, if present, corresponds to the input channels. If you invert S using ISTFT and want the result to have the same number of time samples NT as X, then $(NT - NOVERLAP)/(length(WINDOW) - NOVERLAP)$ must equal an integer.

$S = stft(X,Fs)$ specifies the sample rate of X in Hertz as a positive scalar. This parameter provides time information to the input and only applies when X is a vector or a matrix.

$S = stft(X,Ts)$ specifies Ts as a positive scalar duration corresponding to the sample time of X. This parameter provides time information to the input and applies only when X is a vector or a matrix. The sample rate in this case is calculated as $1/Ts$.

$[S,F,T] = stft(...)$ returns the times at which the stft is evaluated. If a sample rate is provided, T is a vector that contains time values in seconds. If a sample time is provided, then T is a duration array with the same time format as the input. If no time information is provided, the output is a vector in sample numbers. S has a number of rows equal to the length of the frequency vector F and a number of columns equal to the length of the time vector T.

$stft(...)$ with no output arguments plots the magnitude of the `stft`. This syntax does not support multichannel signals.

$X = istft(S)$ returns the inverse Short-Time Fourier transform (`istft`) of S. For single-channel signals, specify S as a matrix with time increasing across the columns and frequency increasing down the rows. For multichannel signals, specify S as a 3-D array with the third dimension corresponding to the channels. S is expected to be two-sided and centered.

$X = istft(S,Fs)$ specifies the sample rate of X in Hertz as a positive scalar. $X = istft(S,Ts)$ specifies Ts as a positive scalar duration corresponding to the sample time of X. The sample rate in this case is calculated as $1/Ts$.

$X = istft(..., 'Window', WINDOW)$ specifies the window used in calculating the `istft`. Perfect time-domain reconstruction requires the istft window to match the window used to generate the STFT. Use the function *ISCOLA* to check a window/overlap combination for constant overlap-add (*COLA*) compliance. *COLA* compliance is a requirement of perfect reconstruction of non-modified spectra. The default is a Hann window of length 128.

A sample MATLAB script to compute the STFT of a signal and reconstruct the original signal using the ISTFT function is appended below.

```
%Generate a Chirp signal sampled at 1 kHz for 1 second.
%The concave quadratic chirp signal has an instantaneous frequency 100 Hz
% at t = 0 and crosses 300 Hz at t = 1 second. It has an initial phase
% equal to 45 degrees. Compute the STFT of the signal using a periodic
% Hamming window of length 256 and an overlap length of 15 samples.
% Plot the original and reconstructed versions.
clear all; close all; clf;
fs = 1e3; t = 0:1/fs:1-1/fs;
x = [chirp(t,100,1,300,'quadratic',45,'concave')];
subplot(311);
stft(x,fs,'Window',hamming(256,'periodic'),'OverlapLength',15);
hold on;
[S,F,T] = stft(x,fs,'Window',hamming(256,'periodic'),'OverlapLength',15);
[xi,ti] = istft(S,fs,'Window',hamming(256,'periodic'),'OverlapLength',15);
```

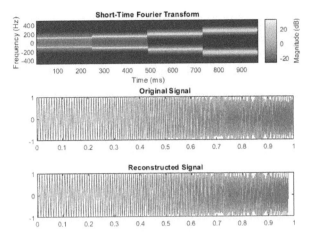

Figure 3.6 Illustration of STFT and ISTFT Functions

```
subplot(312),plot(t,x);
title('Original Signal');
subplot(313),plot(ti,xi);
title('Reconstructed Signal'); hold off;
```

The output of the above MATLAB script is shown in figure 3.6.

3.2.15   WAVELET TRANSFORM AND INVERSE

It was Grossman and Morlet in 1984 who first proposed the idea of a *wavelet*. Note that certain seismic signals can be modeled suitably by combining translations and dilations of a simple, oscillatory function of finite duration called a wavelet.

3.2.15.1   Continuous-Time Wavelets

A real or complex=value continuous-time function $\psi(t)$ satisfying the following properties is called a mother wavelet (or simply a wavelet):

1. The function integrates to zero: $\int_{-\infty}^{\infty} \psi(t)dt = 0$.

2. It is square integrable, or, equivalently, has finite energy: $\int_{-\infty}^{\infty} |\psi(t)|^2 dt < 0$.

3. The admissibility condition: $C = \int_{-\infty}^{\infty} \frac{|\psi(\omega)|^2}{|\omega|} d\omega$, such that $0 < C < \infty$.

As an example, a *Morlet Wavelet*, which is constructed by modulating a sinusoidal function by a Gaussian function is a wavelet of infinite duration. But, most of the

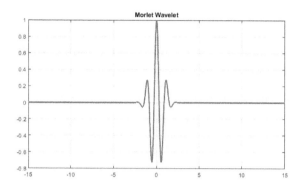

Figure 3.7   The Morlet Wavelet Function

energy in this wavelet is confined to a finite interval. The real-value Morlet wavelet
is given by:

$$\psi(t) = e^{-t^2} \cos\left(\pi\sqrt{\frac{2}{ln2}}t\right) \tag{3.2.43}$$

The Morlet Wavelet is shown in figure 3.7. The MATLAB/GNU Octave code to
generate the same is given below:

```
%Illustration of Morlet Wavelet..
clear all; close all; clf;
t=[-15:0.01:15];
mor=exp(-t.^2).*cos(pi*sqrt(2/log(2)).*t);
plot(t,mor,'LineWidth',2);
title("Morlet Wavelet");grid;
```

3.2.15.2   Definition of the Continuous-Time Wavelet Transform (CWT)

Let $f(t)$ be any square-integrable function. The CWT of $f(t)$ w.r.t. a wavelet $\psi(t)$ is
defined as

$$W(a,b) = \int\limits_{-\infty}^{\infty} f(t)\frac{1}{\sqrt{|a|}}\psi^*\left(\frac{t-b}{a}\right)dt \tag{3.2.44}$$

where a and b are real and $*$ denotes complex conjugation. Now, by defining
$\psi_{a,b}(t) = \frac{1}{\sqrt{|a|}}\psi\left(\frac{t-b}{a}\right)$, we can write:

$$W(a,b) = \int\limits_{-\infty}^{\infty} f(t)\psi_{a,b}^*(t)dt \tag{3.2.45}$$

3.2.15.3 Inverse Wavelet Transform

If the mother wavelet satisfies the admissibility condition, then:

$$f(t) = \frac{1}{C} \int_{a=-\infty}^{\infty} \int_{b=-\infty}^{\infty} \frac{1}{|a|^2} W(a,b) \psi_{a,b}(t) da\, db \qquad (3.2.46)$$

where $C = \int_{-\infty}^{\infty} \frac{|\psi(\omega)|^2}{|\omega|} d\omega$, such that $0 < C < \infty$

3.2.15.4 Applications of Wavelet Transforms

1. Wavelets are suited for communication systems due to its potential to handle non stationary behavior and to segregate information into uncorrelated segments.
2. Orthogonal Frequency Division Multiplexing (OFDM) uses multicarrier technique by dividing the spectrum into many subcarriers and each subcarrier gets modulated by low data rate. Wavelet packet transforms can be effectively used in OFDM systems.
3. Biomedical signal processing is a promising field for future research. Wavelet transform can be used effectively in combination with artificial intelligence to provide solution to many problems. Biomedical signals are generally one-dimensional time series data (Electro Cardiogram- ECG, electroencephalogram -EEG) or an image (X-ray, ultrasound scan, MRI). Accordingly, a 1D or 2D wavelet transform can be used to process the signal.
4. Wavelets can be effectively used for abnormality detection in biomedical images. In [13] wavelet decomposition coefficients are used to extract features by calculating 2-level *Haar Wavelet* transform and extract mean, standard deviation and energy of the transform coefficients as features for extraction of abnormal areas in image.

3.2.15.5 MATLAB Function for Computation of Wavelet Transform

CFS = wt(FB,X) returns the continuous wavelet transform (CWT) coefficients of the signal X, using the CWT filter bank, *FB*. X is a double- or single-precision real- or complex-valued vector. X must have at least four samples. If X is real-valued, *CFS* is a 2-D matrix where each row corresponds to one scale. The column size of *CFS* is equal to the length of X. If X is complex-valued, *CFS* is a 3-D matrix, where the first page is the CWT for the positive scales (analytic part or counterclockwise component) and the second page is the CWT for the negative scales (anti-analytic part or clockwise component).

[CFS,F] = wt(FB,X) returns the frequencies, F, corresponding to the scales (rows) of *CFS* if the 'SamplingPeriod' property is not specified in the CWTFILTERBANK, *FB*. If we do not specify a sampling frequency, F is in cycles/sample.

[CFS,F,COI] = wt(FB,X) returns the cone of influence, *COI*, for the CWT. *COI* is in the same units as F. If the input X is complex, *COI* applies to both pages of *CFS*.

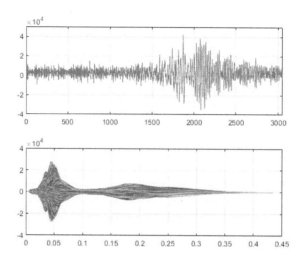

Figure 3.8   The Wavelet Transform of Kobe Earthquake Data

[CFS,P] = *wt(FB,X)* returns the periods, *P*, corresponding to the scales (rows) of
CFS if you specify a sampling period in the CWTFILTERBANK, *FB*. *P* has the
same units and format as the duration scalar sampling period.
[CFS,P,COI] = *wt(FB,X)* returns the cone of influence in periods for the CWT. *COI*
is an array of durations with the same Format property as the sampling period. If the
input *X* is complex, *COI* applies to both pages of *CFS*.

A sample MATLAB script to compute the CWT of the Kobe earthquake data is
given below. The output generated is shown in figure 3.8

```
% Obtain the continuous wavelet transform of the
% Kobe earthquake data.
clear all; close all; clf;
load kobe;
fb = cwtfilterbank('SignalLength',numel(kobe));
[cfs,f] = wt(fb,kobe);
subplot(211),plot(kobe);
axis([0 3050 -4e4 5e4]);grid;
subplot(212),plot(f,cfs);grid;
```

3.2.15.6   Discrete Wavelet Transform (DWT) and IDWT

We will now consider the MATLAB function to compute the Discrete Wavelet Trans-
form and its inverse.
[CA,CD] = *dwt(x,'wname')* returns the single-level discrete wavelet transform
(DWT) of the vector *x* using the wavelet specified by '*wname*'. The wavelet must

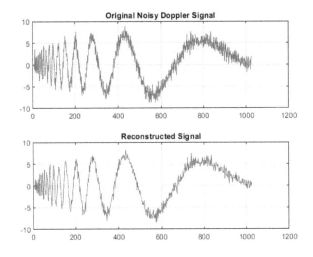

Figure 3.9   The DWT and IDWT of a Noisy Doppler Signal

be recognized by wavemngr function in MATLAB. dwt returns the approximation coefficients vector *CA* and detailed coefficients vector *CD* of the DWT.

x = *idwt(CA,CD, 'wname')* returns the single-level reconstructed approximation coefficients vector *x* based on approximation and detail coefficients vectors *CA* and *CD*, and using the wavelet 'wname'.

Now, we will develop a MATLAB script to compute the single-level DWT of the noisy Doppler signal using the sym4 wavelet. Then, we will reconstruct a smoothed version of the signal using the approximation coefficients, and plot and compare with the original signal.

```
% MATLAB Script to compute the DWT and IDWT.
load noisdopp;
[CA,CD] = dwt(noisdopp,'sym4');
xrec = idwt(CA,zeros(size(CA)),'sym4');
subplot(211),plot(noisdopp);grid;
title('Original Noisy Doppler Signal');
subplot(212),plot(xrec);grid;
title('Reconstructed Signal');
```

The output generated is illustrated in figure 3.9

3.3   SIGNAL THEORY

A Signal is a quantity which convey information about the state of a physical system. It is the transducer that converts the physical quantity which is being monitored or observed into an electrical signal. For example, in a voice-to-voice communication system (a PA system), the microphone and the loudspeaker act as the transducers. Mathematically, a signal S is represented as a set of all elements x, that satisfies a property P. It can be written as:

$$S = \{x; P\} \tag{3.3.47}$$

Hence, S is the set of all elements x such that the property P is true.

3.3.0.1 Mappings and Functionals

A mapping is simply the rule by which elements in one set, say S_1, are assigned to elements in another set, say S_2. Symbolically, the mapping is denoted as $f : S1 \rightarrow S_2$, which is a compact notation for

$$y = f(x); \quad x \in S_1 \ \text{and} \ y \in S_2 \tag{3.3.48}$$

The element y in S_2 is called the image of x under the mapping f. The set S_1 is the domain of the mapping, and the set S_2 is the range of the mapping. The relation f that maps S_1 to S_2 is called the functional.[2] Note that, all transforms, are mappings that are used extensively in signal analysis.

3.3.0.2 Time-Frequency Duality

There is a one-to-one correspondence between the sets of square-integrable functions connected by the Fourier Transforms. Also note the essentially symmetric nature of the Fourier Transform mapping and its inverse. Due to this, we find that for every relationship involving time functions, there is a corresponding dual relationship involving Fourier Transforms. This time-frequency duality property exhibited by time functions and their Fourier Transforms is frequently exploited in signal analysis problems.

3.3.1 OUTPUT OF AN LTI SYSTEM

The input-output relationship of a Linear-Time Invariant (LTI) system is governed by the appropriate transfer function. This follows from the fact that, in an LTI system, the output signal is related to the input signal and impulse response of the LTI system by the convolution formula. That is:

$$y(t) = x(t) * h(t) = \int_{\tau=-\infty}^{\infty} x(\tau)h(t-\tau)d\tau$$

Note that if the Laplace Transform of a continuous time LTI system input is $X(s)$, and transform of its unit impulse response, $h(t)$ is $H(s)$, then its transform of the output function $f(t)$ denoted by $Y(s)$ is expressed as:

$$Y(s) = X(s)H(s)$$

---

[2]Sometimes the terms mappings and functions are used synonymously.

Now, we can easily arrive at the output function $y(t)$ by taking the Inverse Laplace Transform. Hence, we have:

$$y(t) = \mathscr{L}\mathscr{T}^{-1}\{Y(s)\} = \mathscr{L}\mathscr{T}^{-1}\{X(s).H(s)\}$$

This technique is applicable to other appropriate transforms as well.

3.3.2 OUTPUT OF A NLTV SYSTEM

In the case of NLTV systems, the input and the output are not governed by the convolution formula. Hence we can not apply the transfer function technique to arrive at the expression for the output function.

3.4 CONCLUDING REMARKS

In this chapter, we discussed Signal Processing or processing (transformation) of signals in `frequency domain`. It may be noted that signal processing in frequency domain has certain advantages compared to that in time domain. Transforms are discussed as a viable means of converting signals in time domain to frequency domain. We considered various *transforms* like Laplace Transform, Fourier Transform, z-transform, Hilbert Transform, Hough Transform, Discrete Cosine Transform (DCT), Short Term Fourier Transform (STFT), Wavelet Transform, etc., that enabled signal processing in frequency domain. We considered the mathematical formulations of all the above transforms and their inverses. Time domain processing of signals is not really popular due to many reasons, including the scarcity of appropriate hardware. However, recently time domain processing has also gained some attention from practicing engineers and scientists.

FURTHER READING

1. K.C. Raveendranathan, *Communication Systems Modelling and Simulation Using MATLAB and Simulink*, 1st Edition, Universities Press, Hyderabad, 2011.
2. B.P. Lathi, *Signal Processing & Linear Systems*, Oxford University Press, 2008.
3. P A Cook, *Nonlinear Dynamical Systems*, 2nd Edition, Prentice Hall (UK), 1994.
4. Rafael C. Gonzalez, Richard E. Woods, and Steven L. Eddins, Digital Image Processing Using MATLAB, 3rd Edition, Gatesmark Publishing, 2020.
5. James W. Cooley and John W. Tukey, An Algorithm for the Machine Computation of the Complex Fourier Series, *Mathematics of Computation*, Vol. 19, April 1965, pp. 297–301.
6. N. Ahmed, T. Natarajan, and K.R. Rao, Discrete Cosine Transform, *IEEE Transactions on Computers*, January 1974, pp.90–93.
7. Raghuveer M. Rao and Ajit S. Bopardikar, *Wavelet Transforms: Introduction to Theory and Applications*, Pearson Education, 2000.
8. L.E. Franks, *Signal Theory*, Dowden and Culver, 1981.
9. Alan V. Oppenheim, Ronald W. Schafer, and John R. Buck, *Discrete-Time Signal Processing*, 2nd Edition, Pearson, 2011.

10. Edward B. Magrab, *An Engineer's Guide to Mathematica*, John Wiley, 2014.
11. Stephen Wolfram, *An Elementary Introduction to the Wolfram Language*, 3rd Edition, 2023.
12. Hans-Petter Halvorsen, *Python Programming*, 2020.
13. Hans-Petter Halvorsen, *Python for Science and Engineering*, 2020.

EXERCISES

1. Find the *Inverse Laplace Transform* of:
   a. $\frac{5s-4}{s^2-s-12}$.
   b. $\frac{2s^2+5}{s^2+3s+2}$.
   c. $\frac{5(s+32)}{s(s^2+12s+32)}$.
   d. $\frac{6s+10}{(s+1)(s+2)^2}$.
   e. $\frac{s+4+4e^{-2s}}{(s+1)(s+2)}$.

2. Using the *time convolution property* of the Laplace Transform, find:

$$x(t) = e^{-at}u(t) * e^{-bt}u(t)$$

3. State whether the following signals are *periodic* or *aperiodic*:
   a. $2\cos t + 3\cos 3t$
   b. $4 + 12\sin 5t + 5\cos 8t$
   c. $5\cos\sqrt{5}t - 5\cos 2t$
   d. $(4\cos 2t + \cos 5t)^2$

4. Consider a system with impulse response $h[n] = u[n] - u[n-5]$ and input sequence, $x[n] = a^n u[n]$. Calculate the output.

5. Find the *Inverse Fourier Transform* of $\delta(\omega - \omega_0)$.

6. Find the z-Transform of the sequence $x[n] = \begin{cases} n, & 0 \le n \le 9 \\ 10, & n \ge 10. \end{cases}$

7. Obtain the *Fourier Transform* of the *sign function* given by:

$$sgn(t) = \begin{cases} 1 & t > 0 \\ -1 & t < 0 \end{cases}$$

8. Show that if $x(t)$ is an *even function* of t (i.e. $x(t) = x(-t)$), then $X(j\omega) = 2\int_0^\infty x(t)\cos\omega t\, dt$ and if $x(t)$ is an *odd function* of t (i.e. $x(-t) = -x(t)$), then

$$X(j\omega) = -2j\int_0^\infty x(t)\sin\omega t\, dt.$$

9. A signal $x(t)$ can be expressed as the sum of *even* and *odd* components as:

$$x(t) = x_e(t) + x_o(t)$$

Then, if $x(t) \leftrightarrow X(j\omega)$ and if $x(t)$ is real, show that:

$$\begin{aligned} x_e(t) &\leftrightarrow \Re\{X(j\omega)\} \\ x_o(t) &\leftrightarrow j\Im\{X(j\omega)\} \end{aligned} \qquad (3.4.49)$$

10. A *causal* LTI system has impulse response $h[n]$ for which $H[z] = \frac{1+z^{-1}}{(1-0.5z^{-1})(1+0.25z^{-1})}$.

    a. Find the impulse response $h[n]$.

    b. What is the ROC of $H[z]$.

    c. Is the system *stable?* Explain.

    d. Find the z-Transform $X[z]$ of an input sequence $x[n]$ for which the output of the system will be:

$$y[n] = -\frac{1}{5}\left(-\frac{1}{4}\right)^n u[n] - \frac{4}{3}(2)^n u[-n-1].$$

11. For the following pairs of $X[z]$ and $Y[z]$, obtain the *system function*, $H[z]$ and its ROC:

$$X[z] = \frac{1}{1-0.75z^{-1}}, \quad |z| > 0.75$$

$$Y[z] = \frac{1}{1+\frac{2}{3}z^{-1}}, \quad |z| > \frac{2}{3}$$

12. The input sequence to a *causal LTI system* is $x[n] = u[-n-2] + \left(\frac{1}{2}\right)^n u[n]$. The z-Transform of the output of the system is $Y[z] = \frac{-0.5z^{-1}}{(1-0.5z^{-1})(1+z^{-1})}$.

    a. Obtain the z-Transform of the impulse response of the system, $H[z]$.

    b. Find $y[n]$.

    c. What is the ROC for $Y[z]$?

13. Find the sequence $x[n]$ with z-Transform $X[z] = (1+3z)(1+2z^{-1})(1-z^{-1})$.

14. Obtain the *Fourier Transform* of the *unit triangle function*:

$$\Delta(x) = \begin{cases} 0 & |x| \geq \frac{1}{2} \\ 1-2|x| & |x| < \frac{1}{2} \end{cases}$$

15. For an LTI continuous-time system with *transfer function* $H(s) = \frac{2}{s+2}$, find the *zero state response* if the input is:

    a. $e^{-2t}u(t)$.

    b. $e^{-4t}u(t-2)$.

    c. $e^t u(-t)$.

    d. $u(t+2)$.

16. Show that the energy of a *Gaussian Pulse* $x(t) = \frac{1}{\sigma\sqrt{2\pi}}e^{-\frac{t^2}{2\sigma^2}}$ is $\frac{1}{2\sigma\sqrt{\pi}}$.

17. Obtain the z-Transform and the ROC of the *non-causal* sequence $x[n] = -a^n u[-n-2]$. Also, sketch the sequence using MATLAB, Mathematica, GNU Octave and Python.

18. Consider a system with system function as $H(z) = \frac{2z^4+3z^3-4z^2+6z+8}{2z^3+z^2-4z+7}$. Find out whether the system is *causal*.

19. Find the *inverse system* of the system characterized by $y[n] = \beta y[n-1] + x[n]$. (Hint: If the system function of the inverse system is $H^{-1}[z]$, then $H[z] \times H^{-1}[z] = 1$.)

20. Let $x = [1\,2\,3\,4]$ Write the MATLAB/GNU Octave/Mathematica/Python code to get:

$$f[x] = x - \frac{\sin x - \cos x}{\sin x + \cos x}$$

21. The impulse response of a system is $h(t) = \begin{cases} t^2 e^{-t^2} & t > 0 \\ 0 & t < 0 \end{cases}$ Using MATLAB/Octave/Mathematica/Python obtain the response to the system to the input $x(t) = \begin{cases} B & 0 < t < T \\ 0 & otherwise \end{cases}$

22. Using MATLAB/GNU Octave/Mathematica/Python obtain the *spectrum* of a rectangular pulse with duration $250ms$ and amplitude $2V$.

23. Use MATLAB/GNU Octave/Mathematica/Python to plot the function $f(t) = sinc(200t)\sin(2\pi 200t)$.

24. Obtain the output of an LTI system with impulse response $h(t) = \delta(t) + \delta'(t)$, where $\delta'(t) = \frac{d\delta(t)}{dt}$ and the input applied is $x(t)$.

25. Obtain the system *differential equation* characterized by the following system function: $H(j\omega) = \frac{4}{4+j\omega}$.

26. Let $X(e^{j\omega}) \leftrightarrow (0.25)^n u[n-1] = x[n]$. Find the sequence corresponding to $Y(e^{j\omega}) = X(e^{j\omega}) * X(e^{j\omega})$.

27. The *unit step response* of an LTI system is $s[n] = u[n-1] - u[n-4]$. Obtain its *unit impulse response*.

28. Check whether the discrete-time system is characterized by the input-output relation $y[n] - \frac{1}{4}y[n-1] - \frac{1}{5}y[n-2] = x[n]$ is *BIBO stable*.

29. Let $F[z] = \frac{z^2(z-2)}{(z^2-2)(z-0.25)}$. Obtain $f[0]$ and $\lim_{n\to\infty} f[n]$.

30. Obtain the *system difference equation* of the *causal* system with *system function* $H[z] = \frac{2z}{z^2+2z+1}$.

31. Find the z-Transform and the ROC of the causal sequence $x[n] = u[n+1] - u[n-1]$.

32. Let the input signal $x(t)$ is $12\cos(2\pi ft) + 14\sin(4\pi ft)$ where $f = 4kHz$. The signal is passed through an LTI system with *impulse response* $h(t) = 1$, *for* $t \in [0, 1ms]$. Obtain the following using MATLAB/GNU Octave/Mathematica/Python:
    a. The Fourier Transform of $x(t)$.
    b. The *power spectrum* of $x(t)$.
    c. The *transfer function* of the LTI system and a qualitative plot of the same.
    d. The output signal, $y(t)$.
    e. The *average power* of the output signal $y(t)$.

33. Show that $\int\limits_{-\infty}^{\infty} sinc^2(kt)\,dt = \frac{\pi}{k}$.

34. Show that for *real, Fourier Transformable signals*, $x_1(t)$ and $x_2(t)$,

$$\int\limits_{-\infty}^{\infty} x_1(t)x_2(t)dt = \frac{1}{2\pi}\int\limits_{-\infty}^{\infty} F_1(-j\omega)F_2(j\omega)d\omega = \frac{1}{2\pi}\int\limits_{-\infty}^{\infty} F_1(j\omega)F_2(-j\omega)d\omega$$

4 Introduction to Communication Systems

A communication system is an *entity* that enables transfer of *information* from one point to another. The origin of information is called the *information source* and the end point is called the *information sink*. There are several other subsystems or blocks in between the source and the sink, each serving a unique purpose. In this chapter, we focus on the various subsystems of a typical communication system. The signal or information passes from source to destination through what is called *channel*, which represents the medium that carries the signal around. Communication systems can be either *wired* or *wireless*, depending on whether there is a physical connection between the subsystems involved or not. One can also classify communication systems as *baseband* or not. An example for baseband communication system (in which there is no modulator or demodulator) is a Public Addressing system (PA system). Almost all practical communication systems contain the modulator and the demodulator, due to some specific advantages. This will become more evident later. We will consider the basic building blocks of communication systems in the following sections.

4.1 INTRODUCTION

Figure 4.1 illustrates a typical communication system. The subsystems which are part of the *transmitter* are the *input transducer* (which converts the physical quantity to an electrical voltage or current), *source encoder*, *channel encoder*, and *modulator.* The *receiver* consists of the *demodulator (or detector)*, the *channel decoder*, and the *source decoder.* It may be noted that whatever one does to the signal at the transmitter has to be undone at the receiver. Also, the receiver must contain an output transducer which converts the electrical signal back to the physical signal such as as acoustic waves. The channel is invariably noisy, meaning a certain amount of noise signals are getting added to signal at the channel. Hence, the receiver should also contain a suitable filter as one of its building block. Modulation and demodulation are two important steps in information transmission and reception. Therefore, modulators and demodulators deserve a lot of attention from the communication engineers. In the succeeding sections, we will consider a number of different modulation and demodulation schemes.

4.2 OVERVIEW OF COMMUNICATION SYSTEMS

As mentioned earlier, communication systems can be either wired or wireless. A hybrid form is also possible. The channel can be free space or a physical medium like an electrical cable or an optical fiber cable. If the channel is wireless, transmitter and

DOI: 10.1201/9781003527589-4

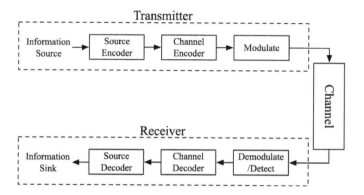

Figure 4.1   The Block Diagram of a Typical Communication System

receiver antennas be come part of the communication system. Impedance matching is an important criterion that has to be met while interconnecting various subsystems. Antenna matching unit (AMU) is used to couple the power amplifiers following the modulator to the antenna, so that efficient transmission of signals is made possible. This is so because we have to match the impedance of the antenna to that of the intrinsic impedance of free-space, which is the channel [1].

4.2.1   SOURCE ENCODER

The signal source is highly redundant. Hence, we employ the source encoder to reduce the redundancy among source symbols, before they are applied to the channel encoder. In other words, the source encoder is responsible for compressing the input information sequence to represent it with less redundancy. The compressed data is passed to the channel encoder. Source encoding aims to convert information waveforms (text, audio, image, video, etc.) into bits, the universal chunk of information in the digital world.

4.2.2   CHANNEL ENCODER

The channel encoder introduces some redundancy in the binary information sequence that can be used by the channel decoder at the receiver to overcome the effects of noise and interference encountered by the signal while in transit through the communication channel. It may be emphasized that the goal of source coding is to represent a source with the lowest possible rate to achieve a particular distortion, or with the lowest possible distortion at a given rate. Channel coding, on the other hand, adds redundancy to quantized source information to recover channel errors.

---

[1]The intrinsic impedance of free space is calculated as $120\pi \simeq 376.98\Omega$.

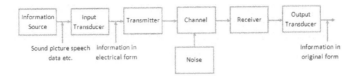

Figure 4.2   The Block Diagram of an Analog Communication System

4.3   ANALOG COMMUNICATION SYSTEMS

An analog communication system is one where the information signal sent from point A to point B can only be described as an analog signal. Example for an analog communication system is the analog Plain Old Telephone System (POTS). Terrestrial analog Television System is another example. In analog communication system, the information is transferred as analog signals between transmitter and receiver. The block diagram of an analog communication system is shown in figure 4.2. Of late, analog communication systems are gaining less popularity. Now, most of the communication systems are *digital* and *wireless*.

4.3.1   NEED FOR MODULATION & DEMODULATION

In baseband communication, the information signal is transmitted as such without any modulation. Modulation is a process where a suitable parameter like amplitude, frequency, or phase of a carrier wave is modified according to the amplitude of the information signal. The reasons for modulation are the following:

1. Baseband transmission is highly inefficient in information transmission. The signal coverage will be limited without modulation.
2. The antenna length has to be inappropriately high for base band transmission. The typical antenna length has to a tenth of the wavelength of the signal being transmitted, for efficient radiation. Hence, by using a carrier wave of higher frequency, we can effectively reduce the antenna length.
3. By choosing a high-frequency carrier wave, we can introduce noise immunity in communication. Note that certain spectrum of frequencies are less susceptible to noise than others.
4. We can effectively incorporate multiplexing (i.e. sending more signals using the same channel) by modulation, with out signal interference.

4.4   AMPLITUDE MODULATION (AM)

In Amplitude Modulation, the amplitude of a higher frequency carrier wave is modified or changed as per the instantaneous amplitude of the information signal (modulating signal) before it is transmitted. In practice, the carrier frequency must be 50 to 100 times the frequency of the highest frequency present in the modulating signal. The amplitude of the carrier signal is so chosen that it is always greater than the amplitude of the modulating signal.

4.4.1   MATHEMATICAL MODEL OF AM

The carrier wave is sinusoidal wave of amplitude E_c and frequency, f_c and the modulating signal is of amplitude E_m and frequency f_m. Thus the instantaneous values of the carrier and modulating signals are:

$$
\begin{aligned}
e_c &= E_c \sin(2\pi f_c t) \\
e_m &= E_m \sin(2\pi f_m t)
\end{aligned}
$$

The instantaneous value of the modulated, AM signal is:

$$
\begin{aligned}
e_{AM} &= [E_c + E_m \sin(2\pi_m t)] \sin(2\pi f_c t) \\
&= E_c \left[1 + \frac{E_m}{E_c} \sin(2\pi f_m t)\right] \sin(2\pi f_c t) \\
&= E_c [1 + m_a \sin(2\pi f_m t)] \sin(2\pi f_c t), \;\; \text{where } m_a = \frac{E_m}{E_c} \quad (4.4.1)
\end{aligned}
$$

Note that $f_c \gg f_m$. We call $m_a = \frac{E_m}{E_c}$ as the *modulation index* or the *modulation depth*. We can further simplify equation 4.4.1 as:

$$
\begin{aligned}
e_{AM} &= E_c \sin(2\pi f_c t) + m_a \sin(2\pi f_m t) \sin(2\pi f_c t) \\
&= E_c \sin(2\pi f_c t) + \frac{m_a E_c}{2} \cos[2\pi(f_c - f_m)t] - \frac{m_a E_c}{2} \cos[2\pi(f_c + f_m)t]
\end{aligned}
$$
$$(4.4.2)$$

We can conclude that the amplitude-modulated output wave form contains the carrier frequency as well as two frequencies $f_c - f_m$ and $f_c + f_m$ known as the *lower* and *upper side band* frequencies, respectively.

4.4.2   SIMULATION OF AMPLITUDE MODULATION

We can simulate AM generation in MATLAB/GNU Octave either by explicitly implementing equation 4.4.1 or using the *modulate(.)* or *ammod(.)* functions. A MATLAB script to simulate the generation of AM wave by explicitly implementing the equation 4.4.1 is appended below.

```
% MATLAB script to implement AM A3E generation..
clear all; close all; clf;
t=linspace(0,0.1,1e4);
fc=1e4; fm=400;
Ec=10;Em=8; ma=Em/Ec;
wc=2*pi*fc*t;
wm=2*pi*fm*t;
```

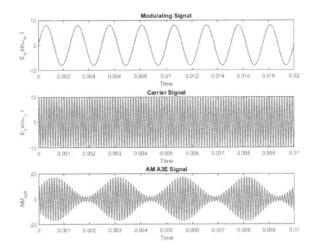

Figure 4.3   Output of A3E AM Wave Generation

```
ec=Ec*sin(wc);
em=Em*sin(wm);
eAM=Ec*(1+ma*sin(wm)).*sin(wc);
subplot(311),plot(t(1:2000),em(1:2000),LineWidth=1.5);
xlabel('Time'); ylabel('E_{m}sin\omega_{m} t');
title('Modulating Signal');
subplot(312),plot(t(1:1000),ec(1:1000),LineWidth=1.5);
xlabel('Time'); ylabel('E_{c}sin\omega_{c} t');
title('Carrier Signal');
subplot(313),plot(t(1:1000),eAM(1:1000),LineWidth=1.5);
xlabel('Time'); ylabel('AM_{out}');
title('AM A3E Signal');
```

Output of the simulation is shown in figure 4.3. Simulation of amplitude modulation using the modulate(.) and ammod(,) functions are enlisted in the MATLAB script given below.

```
% MATLAB script to implement AM A3E generation
% using ammod and modulate functions..
clear all; close all; clf;
t=linspace(0,0.1,1e4);
fc=1e4; fm=400;
fs=4e4; % Sampling frequency.
Ec=10;Em=8; ma=Em/Ec;
wc=2*pi*fc*t;
wm=2*pi*fm*t;
```

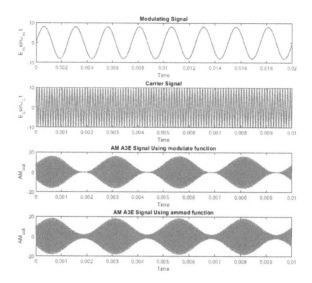

Figure 4.4   Output of A3E AM Wave Generation Using modulate and ammod Functions

```
ec=Ec*sin(wc);
em=Em*sin(wm);
eAM1=modulate(em,fc,fs,'amdsb-tc');
eAM2=ammod(em,fc,fs,0,Ec);
subplot(411),plot(t(1:2000),em(1:2000),LineWidth=1.5);
xlabel('Time'); ylabel('E_{m}sin\omega_{m} t');
title('Modulating Signal');
subplot(412),plot(t(1:1000),ec(1:1000),LineWidth=1.5);
xlabel('Time'); ylabel('E_{c}sin\omega_{c} t');
title('Carrier Signal');
subplot(413),plot(t(1:1000),eAM1(1:1000),LineWidth=1.5);
xlabel('Time'); ylabel('AM_{out}');
title('AM A3E Signal Using modulate function');
subplot(414),plot(t(1:1000),eAM2(1:1000),LineWidth=1.5);
xlabel('Time'); ylabel('AM_{out}');
title('AM A3E Signal Using ammod function');
```

The output waveforms are shown in figure 4.4.

4.4.2.1   Demodulation of A3E Modulation

There are MATLAB functions to retrieve the modulating signal. For example:
x = demod(y,Fc,Fs,METHOD,OPT) demodulates the modulated signal y with a
carrier frequency Fc and sampling frequency Fs, using the demodulation scheme in

METHOD. OPT is an extra, sometimes optional, parameter whose purpose depends on the demodulation scheme we choose. There is another MATLAB function for demodulation of AM waves: *amdemod(.)*.

4.4.3 VARIANTS OF AM-DSBSC, SSB, AND VSB

Note that if we analyze the spectrum of the amplitude-modulated waveform, we can find that it contains the carrier signal and two sidebands. The carrier does not carry any information on the modulating signal, and hence it is redundant. Therefore, we could remove the carrier signal from the AM and still it could convey the information on the modulating signal. This variant of AM is termed as *Amplitude Modulation-Double Side Band Suppressed Carrier (AM-DSBSC)*. It conserves bandwidth as usually $f_c \gg f_m$. We can further reduce the bandwidth without compromising the information carried by transmitting on one of the sidebands (either the LSB or the USB). This variant is called Single Side Band (SSB) AM. Although, both AM-DSBSC and SSB transmission conserve bandwidth, the signal detection at the receiver side is more complex compared to regular AM.

4.4.3.1 Simulation of DSBSC

We can simulate the generation of DSBSC signals using the *modulate(.)* or *ammod(.)* functions. A MATLAB script file using these two functions is appended below. The output of the simulation is illustrated in figure 4.5.

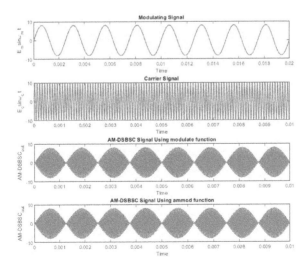

Figure 4.5 Simulation Output of AM-DSBSC Generation Using modulate and ammod Functions

```
% MATLAB script to implement AM-DSBSC generation
% using ammod and modulate functions..
clear all; close all; clf;
t=linspace(0,0.1,1e4);
fc=1e4; fm=400; % Carrier Frequency = 10kHz
% Modulating Frequency = 400Hz
fs=4e4; % Sampling frequency.
Ec=10;Em=8;
wc=2*pi*fc*t;
wm=2*pi*fm*t;
ec=Ec*sin(wc); % Carrier Signal
em=Em*sin(wm);% Modulating Signal
eDSB1=modulate(em,fc,fs,'amdsb-sc');
eDSB2=ammod(em,fc,fs);%AM-DSBSC with zero initial phase
% and zero carrier amplitude.
subplot(411),plot(t(1:2000),em(1:2000),LineWidth=1.5);
xlabel('Time'); ylabel('E_{m}sin\omega_{m} t');
title('Modulating Signal');
subplot(412),plot(t(1:1000),ec(1:1000),LineWidth=1.5);
xlabel('Time'); ylabel('E_{c}sin\omega_{c} t');
title('Carrier Signal');
subplot(413),plot(t(1:1000),eDSB1(1:1000),LineWidth=1.5);
xlabel('Time'); ylabel('AM-DSBSC_{out}');
title('AM-DSBSC Signal Using modulate function');
subplot(414),plot(t(1:1000),eDSB2(1:1000),LineWidth=1.5);
xlabel('Time'); ylabel('AM-DSBSC_{out}');
title('AM-DSBSC Signal Using ammod function');
```

4.4.3.2   Simulation of SSB Generation

To generate SSB, we can multiply e_m by a carrier co-sinusoid of frequency f_c, and add the product to the *Hilbert Transform* of e_m multiplied by a phase-shifted sinusoid of frequency f_c. Hence, the SSB output signal is expressed as:

$$SSB = e_m \times \cos(2\pi f_c t) + \Re \{Hilbert(e_m)\} \times \sin(2\pi f_c t) \qquad (4.4.3)$$

Since $e_m = E_m \sin(2\pi f_m t)$ and $\Re \{Hilbert(e_m)\} = E_m \cos(2\pi f_m t)$, equation 4.4.3 reduces to:

$$
\begin{aligned}
SSB &= E_m \sin(2\pi f_m t) \times \cos(2\pi f_c t) + E_m \cos(2\pi f_m t) \times \sin(2\pi f_c t) \\
&= E_m \{\sin(2\pi f_m t) \times \cos(2\pi f_c t) + \cos(2\pi f_m t) \times \sin(2\pi f_c t)\} \\
&= E_m \sin(2\pi [f_c + f_m]t)
\end{aligned}
$$

Thus, this technique can indeed generate SSB modulation.

The MATLAB functions *modulate*(.) and *ssbmod*(.) can indeed simulate the generation of SSB modulation. The following MATLAB script simulates the generation of SSB. The output of the simulation is shown in figure 4.6

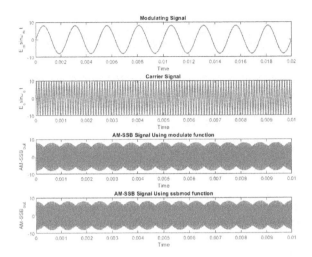

Figure 4.6 Simulation Output of AM-SSB Generation Using modulate and ssbmod Functions

```
% MATLAB script to implement SSB generation
% using  modulate and ssbmod functions..
clear all; close all; clf;
t=linspace(0,0.1,1e4);
fc=1e4; fm=400; % Carrier Frequency = 10kHz
% Modulating Frequency = 400Hz
fs=4e4; % Sampling frequency.
Ec=10;Em=8;
wc=2*pi*fc*t;
wm=2*pi*fm*t;
ec=Ec*sin(wc); % Carrier Signal
em=Em*sin(wm);% Modulating Signal
eSSB1=modulate(em,fc,fs,'amssb');
eSSB2=ssbmod(em,fc,fs,0,'upper');%AM-SSB with zero initial phase.
subplot(411),plot(t(1:2000),em(1:2000),LineWidth=1.5);
xlabel('Time'); ylabel('E_{m}sin\omega_{m} t');
title('Modulating Signal');
subplot(412),plot(t(1:1000),ec(1:1000),LineWidth=1.5);
xlabel('Time'); ylabel('E_{c}sin\omega_{c} t');
title('Carrier Signal');
subplot(413),plot(t(1:1000),eSSB1(1:1000),LineWidth=1.5);
xlabel('Time'); ylabel('AM-SSB_{out}');
title('AM-SSB Signal Using modulate function');
subplot(414),plot(t(1:1000),eSSB2(1:1000),LineWidth=1.5);
```

```
xlabel('Time'); ylabel('AM-SSB_{out}');
title('AM-SSB Signal Using ssbmod function');
```

4.4.3.3 Demodulation of SSB

The MATLAB functions *demod*(.) and *ssbdemod*(.) can simulate the demodulation of SSB modulated signals. For example: x = demod(y,Fc,Fs,METHOD,OPT) demodulates the modulated signal *y* with a carrier frequency Fc and sampling frequency Fs, using the demodulation scheme in *METHOD*. *OPT* is an extra, sometimes optional, parameter whose purpose depends on the demodulation scheme we choose. $Fs > 2 * Fc + BW$, where BW is the bandwidth of the modulated signal. We put 'amssb' to demodulate SSB modulated signals.

The syntax of the *ssbdemod*(.) function is:

x = ssbdemod(y,Fc,Fs,INI_PHASE,NUM,DEN) specifies the numerator and denominator of the lowpass filter to be used in the demodulation. $Fs > 2 * (Fc + BW)$, where BW is the bandwidth of the modulating signal. The demodulation process uses a Butterworth lowpass filter, by default, specified by:

$$[num, den] = butter(5, Fc * 2/Fs)$$

4.4.3.4 Vestigial Side Band (VSB)

Yet another variant of AM transmission is the *Vestigial Side Band (VSB)*, in which a *vestige* or *trace* of the unwanted sideband is transmitted. The VSB scheme is used in video signal transmission of TV signals. The major features of VSB are:

1. VSB is similar to SSB, but it retains a small portion (or vestige) of the undesired sideband to reduce DC distortion.
2. VSB signals are generated by using standard AM or DSBSC modulation, and then passing the modulated signal through a sideband shaping filter.
3. VSB demodulation either uses standard AM or DSBSC demodulation.

4.5 ANGLE MODULATION

In amplitude modulation, the amplitude of the carrier wave is modulated as per the instantaneous value of the modulating signal. The amplitude modulation is a *linear modulation* scheme. We can also modulate the *frequency* or *phase angle* of the carrier waveform. The resulting modulations are called *frequency* or *phase* modulation, respectively. Together, they are termed as *angle modulation*. The angle modulation schemes are *non-linear* in nature. We will discuss both of them in detail in the following subsections.

4.5.1 FREQUENCY MODULATION (FM)

Frequency modulation of a carrier wave occurs when its frequency is varied as per the instantaneous value of the modulating signal. The amplitude of the carrier wave remains constant during this process. The amplitude of the modulating signal will cause the carrier frequency to deviate from its central frequency, i.e., the frequency without any modulation, by a certain amount. As per the regulations of the Federal Communications Commission (FCC), the maximum frequency deviation allowed is $75kHz$. The FCC has allotted the $88 - 108MHz$ band for FM radio broadcast. The ratio of the shift in the carrier frequency from the resting point to the maximum frequency of the modulating signal is called the *deviation ratio*. The FCC has specified the maximum value for deviation ratio as 5. Mathematically:

$$Deviation\ Ratio = \frac{f_{dev}(max)}{f_{AF}(max)}$$

If the maximum frequency deviation of the carrier is $f_{dev}(max)$, and the frequency of the corresponding modulating signal is f_{AF}, then the *modulation index* of the FM waveform is defined as:

$$Modulation\ Index = \beta = \frac{f_{dev}(max)}{f_{AF}}$$

4.5.1.1   The Frequency Spectrum of FM

Assume that the carrier wave in FM is $e_c = E_c \cos \omega_c t$ and the modulating signal by $e_m = E_m \cos \omega_m t$. For simplicity, we assume that there is only one frequency component in the modulating signal. We will remove this restriction later on. The instantaneous value of carrier frequency on modulation, f can be given as:

$$f = f_c + kE_m \cos \omega_m t \qquad (4.5.4)$$

where,

$$
\begin{aligned}
f_c &= \quad unmodulated\ carrier\ frequency \\
k &= \quad proportionality\ constant \\
E_m \cos \omega_m t &= \quad instantaneous\ modulating\ signal.
\end{aligned}
$$

The maximum deviation for the carrier occurs when the cosine term has its maximum value, ± 1. Now, the instantaneous frequency of the modulated carrier will be given by equation 4.5.4. Hence the maximum deviation in frequency, δ will be given by:

$$\delta = \pm kE_m \qquad (4.5.5)$$

The instantaneous amplitude of the FM signal is given by:

$$e = E_c \cos \theta \qquad (4.5.6)$$

Now, we have:

$$\omega = 2\pi f = \omega_c + 2\pi kE_m \cos \omega_m t$$

$$\theta = \int \omega dt = \int (\omega_c + 2\pi kE_m \cos \omega_m t)dt = \left(\omega_c t + \frac{2\pi kE_m \sin \omega_m t}{\omega_m} \right)$$

$$= \omega_c t + \frac{kE_m \sin \omega_m t}{f_m} = \omega_c t + \frac{\delta}{f_m} \sin \omega_m t. \qquad (4.5.7)$$

Substituting the value of θ in equation 4.5.6, we get the instantaneous value of FM signal. Therefore:

$$e = E_c \cos\left(\omega_c t + \frac{\delta}{f_m} \sin \omega_m t\right) \qquad (4.5.8)$$

The *modulation index* for FM, β is defined as:

$$\beta = \frac{Maximum\ Frequency\ Deviation}{Modulating\ Frequency} = \frac{\delta}{f_m} \qquad (4.5.9)$$

Substituting the value of modulation index in equation 4.5.8 we get:

$$e = E_c \cos\left(\omega_c t + \beta \sin \omega_m t\right) \qquad (4.5.10)$$

It can be shown that for a sinusoidal modulating signal, assuming that e represents the instantaneous voltage impressed across a 1Ω resistor, the average power over a cycle of the modulating frequency is given by:

$$\frac{1}{T} \int_0^T e^2\, dt = \frac{E_c^2}{T} \int_0^T \cos^2(\omega_c t + \beta \sin \omega_m t)\, dt \qquad (4.5.11)$$

where $T = \frac{1}{f_m}$. This expression may be rewritten as:

$$\frac{E_c^2}{T} \int_0^T \frac{1 + \cos(2\omega_c t + 2\beta \sin \omega_m t)}{2}\, dt$$

The second part of the integral is zero since it is periodic in T. We assume that $\omega_m = 2\pi/T \ll \omega_c$. The first term gives $\frac{E_c^2}{2}$. Although shown only for a sinusoidal modulating signal, the above result is true for any modulating signal whose highest frequency component $B\ Hz$ is small compared to the carrier frequency f_c.

We will now try to determine the frequency components and their amplitudes for the frequency-modulated signal of arbitrary β and sinusoidal modulating signal. Note that we can write equation 4.5.10 as:

$$e = E_c \left[\cos \omega_c t \cos(\beta \sin \omega_m t) - \sin \omega_c t \sin(\beta\ sin\ \omega_m t)\right] \qquad (4.5.12)$$

We can expand both $\cos(\beta \sin \omega_m t)$ and $\sin(\beta \sin \omega_m t)$ in their respective Fourier series. Both series may be found simultaneously by considering the periodic complex exponential:

$$v(t) = \exp(j\beta \sin \omega_m t), \quad -\frac{T}{2} < t < \frac{T}{2} \qquad (4.5.13)$$

The real part of this function gives us our cosine function; the imaginary part, the sine function. If we expand this exponential in its Fourier series, we can expect to get a real part consisting of even harmonics of ω_m and an imaginary part consisting of the odd harmonics. By equating reals and imaginaries, we shall then obtain the desired Fourier expansions for $\cos(\beta\ \omega_m t)$ and $\sin(\beta\ \omega_m t)$, respectively.

The Fourier coefficient of the complex Fourier series for the exponential of equation (4.5.13) is given by:

$$C_n = \int_{-T/2}^{T/2} \exp\left[j(\beta \sin \omega_m t - \omega_n t)\right] dt; \quad \omega_m = \frac{2\pi}{T}, \quad \omega_n = \frac{2\pi n}{T} = n\omega_m \quad (4.5.14)$$

Normalizing this integral by letting $x = \omega_m t$, we get

$$\frac{C_n}{T} = \frac{1}{2\pi} \int_{-\pi}^{\pi} \exp\left[j(\beta \sin x - nx)\right] dx \quad (4.5.15)$$

This integral can be evaluated only as an infinite series and is called the *Bessel function of the first kind* and is denoted by the symbol $J_n(\beta)$.

Specifically, we can write:

$$J_n(\beta) = \frac{1}{2\pi} \int_{-\pi}^{\pi} \exp\left[j(\beta \sin x - nx)\right] dx \quad (4.5.16)$$

so that $C_n = T J_n(\beta)$.

For $n = 0$, we get $C_0 = T J_0(\beta)$, the dc component of the Fourier series representation of the periodic complex exponential of equation 4.5.13. Increasing values of n give the corresponding Fourier coefficients for the higher frequency terms of the Fourier series. The spectrum of the complex exponential of equation 4.5.13 (and ultimately that of the frequency-modulated signal) will thus be given by the value of the Bessel function and will depend on the parameter β. Hence

$$\exp(j\beta \sin \omega_m t) = \frac{1}{T} \sum_{n=-\infty}^{\infty} C_n \exp(j\omega_n t) = \sum_{n=-\infty}^{\infty} J_n(\beta) \exp(j\omega_n t) \quad \omega_n = n\omega_m$$

$$(4.5.17)$$

This is for $-\frac{T}{2} < t < \frac{T}{2}$.

From equation 4.5.17, for the Fourier series expansion of the complex exponential, we can obtain the desired Fourier series for $\cos(\beta \sin \omega_m t)$ and $\sin(\beta \sin \omega_m t)$. It can be shown, either from the integral definition of $J_n(\beta)$ or from the power series expansion of $J_n(\beta)$, that

$$\begin{aligned} J_n(\beta) &= J_{-n}(\beta) \quad ;n \text{ even} \\ \text{and} \quad j_n(\beta) &= -J_{-n}(\beta) \quad ;n \text{ odd} \end{aligned} \quad (4.5.18)$$

Now writing out the Fourier series term by term, and using equation 4.5.18 to combine the positive and negative terms of equal magnitude of n, we get:

$$\exp(j\beta \sin \omega_m t) = J_0(\beta) + 2\left[J_2(\beta)\cos 2\omega_m t + J_4(\beta)\cos 4\omega_m t + \ldots\right]$$
$$+ 2j\left[J_1(\beta)\sin \omega_m t + J_3(\beta)\sin 3\omega_m t + \ldots\right] \quad (4.5.19)$$

But

$$\exp(j\beta\sin\omega_m t) = \cos(\beta\sin\omega_m t) + j\sin(\beta\omega_m t) \qquad (4.5.20)$$

Equating the real and imaginary terms, we get:

$$\cos(\beta\sin\omega_m t) = J_0(\beta) + 2J_2(\beta)\cos 2\omega_m t + 2J_4(\beta)\cos 4\omega_m t + \ldots$$
$$(4.5.21)$$
$$and \quad \sin(\beta\sin\omega_m t) = 2J_1(\beta)\sin\omega_m t + 2J_3(\beta)\sin 3\omega_m t + \ldots \qquad (4.5.22)$$

Now using the Fourier series expansions for the cosine and sine terms, and then utilizing the trigonometric sum and difference formula, we get:

$$
\begin{aligned}
e \;=\; & E_c\{J_0(\beta)\cos\omega_c t \\
+ \;& J_1(\beta)[\cos(\omega_c + \omega_m)t - \cos(\omega_c - \omega_m)t] \\
+ \;& J_2(\beta)[\cos(\omega_c + 2\omega_m)t + \cos(\omega_c - 2\omega_m)t] \\
+ \;& J_3(\beta)[\cos(\omega_c + 3\omega_m)t - \cos(\omega_c - 3\omega_m)t] \\
+ \;& J_4(\beta)[\cos(\omega_c + 4\omega_m)t + \cos(\omega_c - 4\omega_m)t]\ldots\}
\end{aligned}
\qquad (4.5.23)
$$

We can make the following observations on the frequency-modulated signal:

1. Unlike in AM, *FM signal contains an infinite number of sidebands*, as well as the carrier. They are separated from the carrier by $\pm f_m$, $\pm 2f_m$, $\pm 3f_m, \ldots$, and thus have a recurrence frequency of f_m.
2. The $J_n(\beta)$ coefficients eventually decrease in value as n increases.
3. The *modulation index* determines how many sideband components have significant amplitude.
4. The sidebands at equal distance from f_c have equal amplitudes. The $J_n(\beta)$ coefficients have negative values, occasionally, signifying a 180° phase change for that pair of sidebands.
5. In FM the total transmitted power remains constant, but with increased depth of modulation, the required bandwidth is increased.
6. In FM the *amplitude of the carrier* component *does not remain constant*. Its $J_n(\beta)$ coefficient is J_0, which is dependent on β.
7. It is possible for the carrier component to disappear completely. This happens for certain values of modulation index, called *eigen values*.

4.5.1.2   Simulation of Frequency Modulation

A sample MATLAB/GNU Octave script to generate and plot FM is given below. The output of the simulation is shown in figure 4.7. MATLAB/GNU Octave also has the `modulate(.)` and `fmmod(.)` functions to generate FM. Similarly, `demod(.)` and `fmdemod(.)` functions can be used to demodulate frequency modulated signals. See MATLAB documentation for details.

```
% MATLAB/GNU Octave Script to generate and plot FM wave..
% fc= 10kHz and fm=800Hz
```

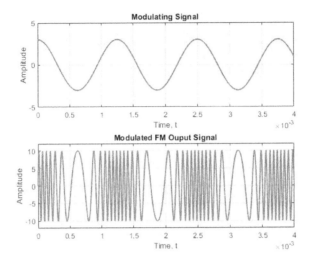

Figure 4.7   Simulation Output of FM Generation

```
% mf=10
clear all; close all; clf;
t=linspace(0,0.004,1e5);
efm=10*sin(2e4*pi*t+10*sin(1600*pi*t));
subplot(211),plot(t,3*cos(1600*pi*t),LineWidth=1.5);
xlabel('Time, t'); ylabel('Amplitude'); grid;
title('Modulating Signal');
axis([0 .004 -5 5]);
subplot(212),plot(t,efm,LineWidth=1.5);
axis([0 .004 -12 12]);
xlabel('Time, t'); ylabel('Amplitude'); grid;
title('Modulated FM Ouput Signal');
```

4.5.2   PHASE MODULATION (PM)

In phase modulation, the phase ϕ of the carrier signal is varied as per the instantaneous value of the modulating signal. The expression for the phase-modulated wave is:

$$e = E_c \sin(\omega_c t + m_p \sin \omega_m t) \qquad (4.5.24)$$

where m_p is the *modulation index* for phase modulation. It must be noted that FM and PM are inseparable, and one can be generated from the other. If the modulating signal $m(t)$ is passed through an integrator and then phase modulated, the resulting waveform will be that of frequency modulation. Similarly, if $m(t)$ is differentiated before passing it to a frequency modulator, the final output will be phase modulated. This is evident from figure 4.8.

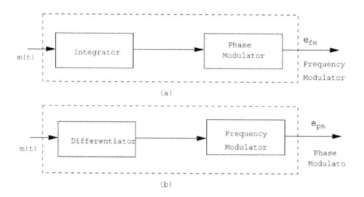

Figure 4.8 Compatibility Between FM and PM

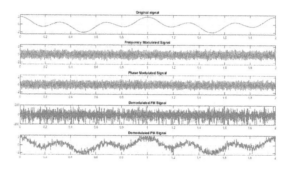

Figure 4.9 Simulation of FM and PM

4.5.2.1   Simulation of FM and PM

The Communications Toolbox in MATLAB/GNU Octave contains functions for the modulation and demodulation of FM and PM. They are the fmmod(.), fmdemod(.), pmmod(.), and pmdemod(.) A MATLAB/GNU Octave script using these functions are appended below. The simulation output is shown in figure 4.9. The modulate and demod functions can also be used for the same purpose.

```
%%% MATLAB/GNU Octave Script to simulate FM & PM.
%% Subsequently  demodulates the both..
clear all; close all; clf;
% Prepare a modulating signal for two seconds,
% at a rate of 1000 samples per second.
Fs =1e3; % Sampling Frequency=1kHz.
t = [0:2*Fs+1]/Fs; % Time points for signals
% Create the modulating signal, as a sum of two cosinusoids.
em = cos(2*pi*t) + cos(6*pi*t);
```

```
Fc = 100; % Carrier frequency in modulation
phasedev = pi/6; % Phase deviation for phase modulation
efm1= fmmod(em,Fc,Fs,phasedev); % Frequency Modulate.
epm1= pmmod(em,Fc,Fs,phasedev); % Phase Modulate.
efm=awgn(efm1,10,'measured'); % Add noise.
epm=awgn(epm1,10,'measured'); % Add noise.
edf = fmdemod(efm,Fc,Fs,phasedev); % Demodulate FM
edp = pmdemod(epm,Fc,Fs,phasedev); % Demodulate FM
% Plot the original, modulated,  and demodulated signals..
subplot(511), plot(t,em,LineWidth=1.5); grid;
axis([0 2 -2.5 2.5]);
title('Original signal');
subplot(512), plot(t,efm,LineWidth=1.5); grid;
axis([0 2 -2.5 2.5]);
title('Frequency Modulated Signal');
subplot(513), plot(t,epm,LineWidth=1.5); grid;
axis([0 2 -2.5 2.5]);
title('Phase Modulated Signal');
subplot(514), plot(t,edf,LineWidth=1.5); grid;
axis([0 2 -200 200]);
title('Demodulated FM Signal');
subplot(515), plot(t,edp,LineWidth=1.5); grid;
axis([0 2 -2.5 2.5]);
title('Demodulated PM Signal');
```

4.6   NOISE IN ANALOG MODULATION SYSTEMS

Now we will consider the noise performance of various analog modulation systems. The concept of *normalized transmission bandwidth* will be introduced to begin with.

$$\text{Normalized Transmission Bandwidth, } \eta = \frac{B_T}{W} \qquad (4.6.25)$$

where B_T is the transmission bandwidth of the modulated signal and W is the bandwidth of the message (baseband). The message bandwidth shows the band of frequencies occupied by the message signal. Table 4.1 lists the normalized transmission bandwidths of various analog modulation schemes.

4.6.1   NOISE IN BASEBAND SYSTEMS

In baseband communication systems, the information (or message) is transmitted without any modulation, and hence the name. The baseband system is suitable for signal transmission over a pair of wires, optical fiber, or coaxial cables. The baseband system is mainly used in *short-haul (short distance)* links.

For a baseband system, the transmitter and receiver are ideal bandpass filters. Refer figure 4.10 for details. The lowpass filters $H_p(f)$ and $H_c(f)$ at the transmitter

Table 4.1

Normalized Transmission Bandwidths, η

System	Transmission Bandwidth, B_T	$\eta = \frac{B_T}{W}$
Baseband	W	$\frac{W}{W} = 1$
AM, A3E	$2W$	$\frac{2W}{W} = 2$
DSBSC	$2W$	$\frac{2W}{W} = 2$
SSB	W	$\frac{W}{W} = 1$
VSB	$W+$	$\frac{W+}{W} = 1+$
FM	$2W(\beta+1)$	$\frac{2W(\beta+1)}{W} = 2(\beta+1)$
PM	$2W(\beta+1)$	$\frac{2W(\beta+1)}{W} = 2(\beta+1)$

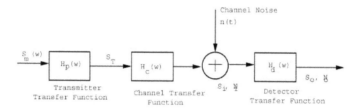

Figure 4.10   Noise Model of Baseband Systems

limits the input signal spectrum to a given bandwidth. The lowpass filter $H_d(f)$ at
the receiver eliminates the out-of-band noise and other channel interference.

The baseband signal is assumed to be a *zero mean, wide-sense stationary random
process* bandlimited to *B Hz*. For a distortion-less channel:

$$S_o = S_i$$

$$N_o = 2\int_0^B S_n(f)\,df$$

where $S_n(f)$ is the *Power Spectral Density (PSD)* of the channel noise. For the case
of white noise: $S_n(f) = \frac{N}{2}$, and

$$N_o = 2\int_0^B \frac{N}{2}\,df = N.B$$

$$\frac{S_o}{N_o} = \frac{S_i}{N.B}$$

We define a parameter Signal-to-Noise Ratio (SNR), γ as

$$\gamma = \frac{S_i}{N.B} = \frac{S_o}{N_o} \qquad\qquad (4.6.26)$$

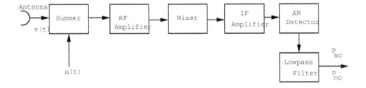

Figure 4.11 The Noise Model of AM Receiver

For voice signals, an SNR of 5 to $10dB$ at the receiver implies a barely intelligible signal. Telephone quality signals have an SNR of 25 to $35dB$, whereas for TV signal reception, an SNR of 45 to $55dB$ is required.

4.6.2 NOISE IN AM RECEIVERS

For AM receivers that carry analog-type message signals, the signal-to-noise ratio is the most commonly used measure of receiver performance. All the noise generated within the receiver can be referred to the receiver input, which makes it easy to compare receiver noise, antenna noise, and received signal. The antenna and receiver gains can be added, so the receiving system can be modeled as shown in figure 4.11.

For determining the signal-to-noise ratio (SNR), the bandwidth of the system is also required. From figure 4.11, it is evident that between the antenna and the detector stage, there is the bandwidth of the RF stages, followed by the bandwidth of the IF stages. Normally, the bandwidth of the IF stages is very much smaller than that of the RF stages, and hence the IF bandwidth will decide the noise that is reaching the detector. From the detector onwards, the bandwidth will be that of the baseband. Let us denote the baseband bandwidth by W and the IF bandwidth by B_{IF}. For AM systems, it may be assumed that $B_{IF} \cong 2W$.

We can deduce that the available noise power output is given by:

$$P_{no} = 2kT_sW$$

where k is the Boltzmann Constant, T_s is the absolute temperature in Kelvins, and W is the bandwidth.

The peak signal voltage at the output is $mE_{c\,max}$, and hence the RMS voltage is mE_c, where E_c is the rms voltage of the unmodulated carrier at the receiver. Therefore the available signal power at the output is:

$$P_{so} = \frac{m^2 E_c^2}{4R_{out}} \tag{4.6.27}$$

Therefore, the output SNR is:

$$\left(\frac{S}{N}\right)_o = \frac{P_{so}}{P_{no}}$$

$$= \frac{m^2 E_c^2}{8R_{out}kT_sW} \tag{4.6.28}$$

The reference noise power at the detector is given by:

$$P_{n\,REF} = kT_s W \tag{4.6.29}$$

The available signal power from a source with an internal resistance R_s is

$$P_R = \frac{E_c^2}{4R_s}\left(1 + \frac{m^2}{2}\right) \tag{4.6.30}$$

Hence, the reference signal-to-noise ratio is given by:

$$\begin{aligned}\left(\frac{S}{N}\right)_{REF} &= \frac{P_R}{P_{n\,REF}} \\ &= \frac{E_c^2(1 + m^2/2)}{4R_s kT_s W}\end{aligned} \tag{4.6.31}$$

A *figure of merit (FOM)* used is the ratio of the above two SNRs.

$$\begin{aligned}FOM_{AM} &= \frac{(S/N)_o}{(S/N)_{REF}} \\ &= \frac{m^2}{2 + m^2} \times \frac{R_s}{R_{out}}\end{aligned} \tag{4.6.32}$$

The higher the figure of merit, better the system. Normally, $R_{out} \cong R_s$ and hence the highest value of attainable FOM is $\frac{1}{3}$, achieved at 100% modulation depth.

4.6.3   NOISE IN COHERENT DSBSC RECEIVERS

In the case of DSBSC receivers, the received signal is of the form

$$e(t) = E_{max} \cos \omega_m t \cos \omega_{IF} t$$

Equation 4.6.28 is applicable to DSBSC case too, with $\frac{E_{max}}{\sqrt{2}}$ replacing $m\,E_c$.

$$\begin{aligned}\left(\frac{S}{N}\right)_o &= \frac{(E_{max}/\sqrt{2})^2}{8R_{out}\,kT_s W} \\ &= \frac{E_{max}^2}{16R_{out}\,kT_s W}\end{aligned}$$

The reference noise power is $P_{n\,REF} = kT_s W$. The RMS value of the received DSBSC signal is $\frac{E_{max}}{2}$. Hence, the available signal power at the input is:

$$\begin{aligned}P_R &= \frac{(E_{max}/2)^2}{4R_s} \\ &= \frac{E_{max}^2}{16R_s}\end{aligned}$$

Therefore, the reference signal-to-noise ratio is:

$$\left(\frac{S}{N}\right)_o = \frac{E_{max}^2}{16R_s k T_s W} \tag{4.6.33}$$

Hence the *figure of merit* is:

$$FOM_{DSBSC} = \frac{(S/N)_o}{(S/N)_{REF}} = \frac{R_s}{R_{out}} \tag{4.6.34}$$

For $R_s = R_{out}$, the FOM_{DSBSC} is unity, which is *thrice* better than that for AM.

4.6.4   NOISE IN SSB RECEIVERS

In SSB, only upper or lower sideband of an AM wave is transmitted. SSB reduces the transmit signal bandwidth by a factor of 2. The transmitted signal can be written in terms $m(t)$ and the Hilbert Transform of $m(t)$ as:

$$s(t) = A_c \left[m(t)\cos(2\pi f_c t) \pm m_h(t)\sin(2\pi f_c t)\right]$$

SSB uses the same demodulator as that of DSBSC. SSB has half the SNR of DSBSC for half the transmit power. Hence there is no SNR gain. It is concluded that SSB receivers have the same noise performance as DSBSC. SSB can introduce significant distortion at DC where the sidebands meet. Hence it is not good for transmission of TV signals.

4.6.5   NOISE IN FM RECEIVERS

Frequency Modulation is much more immune to noise than Amplitude Modulation, and significantly more immune than Phase Modulation. In Frequency Modulation, the SNR improves as the modulating frequency is reduced. Under identical conditions, Frequency Modulation will be $4.75dB$ better in SNR than Phase Modulation. It can be shown that in Frequency Modulation, the noise has a higher effect on the higher modulating frequencies than the lower modulating frequencies.

In Frequency Modulation systems, the amplitude limiters help to reduce impulse-type noise. An important advantage of the Frequency Modulation reception is that an improvement in SNR can be achieved by increasing the frequency deviation. This requires an increased bandwidth, but at least there is an option for trading bandwidth for SNR.

The Power Spectral Density (PSD) is kT_s, and therefore, the available noise power at the detector input for a bandwidth W is $P_{nREF} = kT_s W$. The RMS signal voltage is $E_c = E_{cmax}/\sqrt{2}$, and hence the signal power at the detector input is $P_R = \frac{E_{cmax}^2}{8R_s}$. Hence the reference SNR for the Frequency Modulation system is

$$\left(\frac{S}{N}\right)_{REF} = \frac{P_R}{P_{nREF}} = \frac{E_{cmax}^2}{8R_s k T_s W} \tag{4.6.35}$$

It should be noted that though the reference SNR is decided at the input side of the detector, the bandwidth W is decided by the lowpass filter at the output side.

Consider the signal output from the bandpass filter in the absence of noise, when a sinusoidally modulated carrier is received. The instantaneous frequency is given by:

$$f_i = f_{IF} + \Delta f \cos \omega_m t$$

The Frequency Modulation discriminator converts this to a signal output given by:

$$v_m(t) = \beta \Delta f \cos \omega_m t$$

where β is a constant, which is the frequency-to-voltage coefficient of the discriminator. The RMS voltage at the output is $E_m = \frac{\beta \Delta f}{\sqrt{2}}$, and hence the available power output is:

$$P_{so} = \frac{(\beta \Delta f)^2}{8 R_{out}} \tag{4.6.36}$$

From the definition of PSD, the noise power at the output is:

$$P_{no} = \int_0^W G_v(f) df = \frac{W^3 \beta^2 2k T_s}{3 E_{cmax}^2} \tag{4.6.37}$$

Hence the SNR at the output is:

$$\left(\frac{S}{N}\right)_o = \frac{P_{so}}{P_{no}} = \frac{3 \Delta f^2 E_{cmax}^2}{16 R_{out} k T_s W^3} \tag{4.6.38}$$

Now the *figure of merit* of FM system is

$$FOM_{FM} = \frac{(S/N)_o}{(S/N)_{REF}} = 1.5 \left(\frac{\Delta f}{W}\right)^2 \frac{R_s}{R_{out}} = 1.5 m_f^2 \frac{R_s}{R_{out}} \tag{4.6.39}$$

where m_f is the modulation index calculated for the highest baseband frequency W. Note that this is different from the normal Frequency Modulation index.

4.6.5.1 Carrier-to-Noise Ratio (CNR)

The information on FM signal-to-noise ratio (SNR) is often expressed in another way, using the Carrier-to-Noise Ratio (CNR) as the input parameter at the detector. The CNR is similar to the $(S/N)_{REF}$ except that the total IF bandwidth is used rather than W. So we have:

$$\left(\frac{C}{N}\right) = \frac{E_{cmax}^2}{8 R_s k T_s B_{IF}} \tag{4.6.40}$$

Applying Carson's Rule, $B_{IF} = 2(m_f + 1)W$ gives (valid for $\beta > 6$):

$$\left(\frac{C}{N}\right) = \frac{E_{cmax}^2}{16 R_s k T_s (m_f + 1)W} \tag{4.6.41}$$

4.6.5.2 Processing Gain

The ratio of output SNR to CNR is known as *Processing Gain* and is given by:

$$PG = \frac{(S/N)_o}{(C/N)} = 3m_f^2(m_f + 1)\frac{R_s}{R_{out}} \qquad (4.6.42)$$

It should be noted well that the above result is valid only when the carrier amplitude is much larger than the noise, and the deviation is for a sinusoidally modulated wave. When the carrier level is below the so-called *threshold level*, it is found that the output SNR worsens very rapidly.

4.7 INTRODUCTION TO DIGITAL COMMUNICATION SYSTEMS

The information source produces its output, which is in probabilistic form, as there is no need to convey deterministic source outputs. An input transducer, such as a microphone, converts the source output into a time-varying electrical signal, referred to as the message signal. The transmitter then converts the message signal into a form suitable for transmission through a physical channel, such as a cable. The transmitter generally changes the characteristics of the message signal to match the characteristics of the channel by using a process called modulation. In addition to modulation, other functions, such as filtering and amplification, are also performed by the transmitter.

The communication channel is the physical medium between the transmitter and the receiver, where they are physically separated. No communication channel is ideal, and thus a message signal undergoes various forms of degradation. Sources of degradation may include attenuation, noise, distortion, and interference. As some or all of these degradations are present in a physical channel, an important goal in the design of a communication system is to overcome the effects of such impairments.

The function of the receiver is to extract the message signal from the received signal. The primary function is to perform the process of demodulation, along with a number of peripheral functions, such as amplification and filtering. The complexity of a receiver is generally more significant than that of the transmitter, as a receiver must additionally minimize the effects of the channel degradations. The output transducer, such as a loudspeaker, then converts the receiver output into a signal suitable for the information sink.

The following are highlighted as the major advantages of digital communication systems:

- Design efficiency-Digital communication systems are inherently more efficient than analog counter parts, in exchanging power for bandwidth, the two premium resources in communications. Since an essentially unlimited range of signal conditioning and processing options are available to the designer, effective trade-offs among power, bandwidth, performance, and complexity can be more readily accommodated.
- Availability of versatile hardware.

- Multitude of new and enhanced services.
- Control of quality-A desired distortion level can be initially set and then kept nearly fixed at that value at every step (link) of a digital communication system path. This reconstruction of the digital signal is done by appropriately-spaced regenerative repeaters, which do not allow accumulation of noise and interference.
- Improved security-Digital encryption, unlike analog encryption, can make the transmitted information virtually impossible to decipher.
- Flexibility, compatibility, and switching-Combining various digital signals and digitized analog signals from different users and applications into streams of different speeds and sizes-along with control and signaling information-can be much easier and more efficient.

It may be emphasized that in digital communication systems, the baseband signals are digitized before further processing in the transmitter side. The sampling is to be done a rate not less than *twice* the maximum signal frequency present in the baseband. This rate is known as *Nyquist Rate*. We can retrieve the original analog signal only if the sampling is done at the Nyquist Rate. We will consider the sampling process in the next section.

4.8 PRINCIPLE OF SAMPLING AND ALIASING

Sampling is the process of converting a continuous-time signal into a discrete sequence of data. The *Nyquist-Shannon sampling theorem* depicts the rate at which an analog (continuous time) signal is to be sampled for optimum results.

4.8.1 MATHEMATICAL PROOF OF SAMPLING THEOREM

Let $F(\omega)$ be the spectrum of the signal, $f(t)$. Then, using Inverse Fourier Transform we have:

$$f(t) = \frac{1}{2\pi} \int_{-\infty}^{\infty} F(\omega)e^{j\omega t}d\omega$$

$$= \frac{1}{2\pi} \int_{-2\pi W}^{2\pi W} F(\omega)e^{j\omega t}d\omega$$

since $F(\omega)$ is assumed to be zero outside the band W. Now letting $t = \frac{n}{2W}$ where n is any integer, we can write:

$$f\left(\frac{n}{2W}\right) = \frac{1}{2\pi} \int_{-2\pi W}^{2\pi W} F(\omega)e^{j\omega \frac{n}{2W}}d\omega \qquad (4.8.43)$$

On the left of equation 4.8.43 are values of $f(t)$ at the sampling points. The integral on the right will be recognized as essentially the nth coefficient in a Fourier series expansion of the function $F(\omega)$, taking the interval $-W$ to W as a fundamental period.

This means that the values of the samples $f(n/2W)$ determine the Fourier coeffi-cients in the series expansion of $F(\omega)$. Thus they determine $F(\omega)$, since $F(\omega)$ is zero for frequencies greater than W, and for lower frequencies $F(\omega)$ is determined if its Fourier coefficients are determined. But $F(\omega)$ determines the original function $f(t)$ completely, since a function is determined if its spectrum is known. Therefore the original samples determine the function $f(t)$ completely.

4.8.2 ALIASING

Aliasing arises when a signal is discretely sampled at a rate that is insufficient (less than the Nyquist-Shannon rate) to capture the changes in the signal. Consider the following contexts in which signals are discretely sampled:

- Retinal images are sampled in space by photo receptors.
- Film and video are sampled in time by discrete frames.
- Sound is commonly digitally sampled for recording and communications.

Aliasing may arise in all of these situations if sampling is done improperly. Because the effects of aliasing can be rather disastrous, it is important to understand why aliasing occurs, what its consequences are, and how it may be avoided.

4.9   SOURCE CODING AND CHANNEL CODING

The transmitter consists of the source encoder, channel encoder, and the modulator. The receiver, on the other hand, consists of the demodulator, channel decoder, and the source decoder. At the receiver, the received signal must pass through the inverse subsystems of the corresponding subsystems at the transmitter.

The information may be inherently digital, such as computer data, or analog such as voice. If the information is analog, then the source encoder must first perform *analog-to-digital conversion* to produce a binary stream, and the source decoder must perform a *digital-to-analog conversion* to recover the analog signal. The source en-coder removes the redundancy from the binary stream to make efficient use of the channel. Source coding, a.k.a. data compression leads to conservation of bandwidth, as the spectrum is always a premium resource. The important parameters associated with the source coding are mainly the efficiency of the encoder (the ratio of actual data output rate to the source information rate), and the encoder/decoder complexity.

The channel encoder at the transmitter introduces, in a controlled fashion, redun-dancy. The additional bits are used by the channel decoder at the receiver to overcome the channel-induced errors. The added redundancy serves to enhance the *Bit Error Rate (BER)*, which is the ultimate performance measure of a digital communication system. The important parameters associated with channel coding are mainly the ef-ficiency of the coder (i.e., the ratio of data rate at the input of the encoder to the data rate at its output), error control capability, and the encoder/decoder complexity.

The modulator at the transmitter and the demodulator at the receiver serve as the interfaces to the communication channel. The modulator accepts a sequence of bits, and maps each sequence into a waveform. A sequence may consist of only one or several bits. At the receiver, the demodulator processes the received waveforms, and maps each to the corresponding bit sequence. The important parameters of modulation are the number of bits in a sequence represented by a waveform, the types of waveforms used, the duration of the waveforms, the power level and the bandwidth used, as well as the demodulation complexity.

4.10 PULSE MODULATION SYSTEMS

In this section, we will consider digital modulation of the pulse carrier waveform. We will consider Pulse Amplitude Modulation (PAM), Pulse Width Modulation (PWM) or Pulse Duration Modulation (PDM), Pulse Position Modulation (PPM), Pulse Code Modulation (PCM); and its several variants like Differential Pulse Code Modulation (DPCM), and Adaptive Differential Pulse Code Modulation (ADPCM). When the modulating signal is *discrete*, it can be thought as a sequence of real numbers representing the sample values of an analog waveform. Such type of modulating signals are known as *discrete baseband signals*. One way to send such a signal through a channel is to send a pulse waveform-one pulse is placed at each sampling instant. Such a modulation scheme is known as pulse modulation.

4.10.1 PULSE AMPLITUDE MODULATION (PAM)

The Pulse Amplitude Modulation (PAM) is the transmission of data by varying the amplitudes (voltage or power levels) of the individual pulses in a regularly spaced sequence of electrical or electromagnetic pulses. Pulse Amplitude Modulation (PAM) is the simplest form of pulse modulation. The number of possible pulse amplitudes can be infinite (in the case of analog PAM), but it is usually some power of two so that the resulting output signal can be digital. In certain PAM systems, the amplitude of each pulse is directly proportional to the instantaneous modulating-signal amplitude at the instant the pulse occurs. In other PAM systems, the amplitude of each pulse is inversely proportional to the instantaneous modulating-signal amplitude at the instant the pulse occurs. In still other systems, the intensity of each pulse depends on some characteristic of the modulating signal other than its strength, such as its instantaneous frequency or phase.

4.10.1.1 Generation of PAM

The PAM is generated in much the same manner as analog amplitude modulation. The timing pulses are applied to a pulse amplifier in which the gain is controlled by the modulating waveform. Since these variations in amplitude actually represent the signal, this type of modulation is basically a form of amplitude modulation. The only difference is that the carrier signal is now in the form of pulses. This means that PAM has the same built-in weaknesses as any other amplitude-modulated signal: high susceptibility to noise and interference. The reason for susceptibility to noise is that any interference in the transmission path will either add to or subtract from

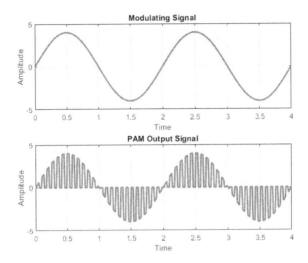

Figure 4.12   The Simulated Pulse Amplitude Modulation Waveform

any voltage already in the circuit (signal voltage). Thus, the amplitude of the signal will be changed. Since the amplitude of the voltage represents the signal, any unwanted change to the signal is considered a *signal distortion*. For this reason, PAM is not often used in practical telecommunication applications. PAM has been superseded by other techniques such as Pulse Position Modulation (PPM) and Pulse Code Modulation (PCM).

While newer technologies are fast making their presence known, it should be noted that pulse amplitude modulation is still useful in the popular Ethernet communication standard. For example, 100BASE-T2 - operating at 100Mb/s - Ethernet medium is using 5-level PAM modulations running at 25 mega pulses/sec over four wires. Later developments include the 100BASE-T medium which raised the bar to 4 wire pairs, running each at 125 mega pulses/sec in order to achieve 1000 Mbps data transfer rates, but still with the same PAM-5 for each pair.

More recently, PAM-12 and PAM-8 have gained consideration in the newly proposed IEEE 802.3an standard for 10GBase-T – 10 gigabyte Ethernet over copper wire.

4.10.1.2   Simulation of PAM in MATLAB

The MATLAB/GNU Octave script to simulate the generation of PAM is given below. The typical output of the simulation is shown in figure 4.12.

```
%% MATLAB/GNU Octave script to simulate PAM...
clear all; close all; clf;
fc=10000; % carrier frequency.
fm=200; % modulating frequency.
```

```
cp=[zeros(1,2),ones(1,4),zeros(1,2)];
n=fc/fm;
t=linspace(0, 4,n*length(cp));
x=4*sin(2*pi*fm*t); %modulating signal..
onx=[];
for i=1:n
    onx=[onx,cp];
end;
pams=x.*onx;
subplot(211), plot(t,x,LineWidth=2);
axis([0 4 -5 5]);
xlabel('Time'); ylabel('Amplitude');
title('Modulating Signal'); grid;
subplot(212), plot(t,pams,LineWidth=2);
axis([0 4 -5 5]);
xlabel('Time'); ylabel('Amplitude');
title('PAM Output Signal');grid;
```

The MATLAB Communications Toolbox function `pammod(.)` and `pamdemod(.)` can also be used to simulate PAM modulation and demodulation, respectively.

4.10.1.3 Demodulation of PAM

Demodulation of PAM is achieved by simply passing the PAM-modulated signal through a low pass filter. The cutoff frequency of the lowpass filter is chosen to be the maximum frequency of the modulating signal.

4.10.2 PULSE WIDTH MODULATION (PWM)

In Pulse Width Modulation, the instantaneous width of a pulse carrier is varied in accordance with the modulating signal. Pulse width modulation is used when a digital system needs to control a system that expects an analog signal of varying amplitude. A typical example is a 12 V motor: the speed of the rotor can be regulated by changing the voltage from low (0 V) to high (12 V). At 12V the motor will go at full speed. The alternative is to pass the rotor always 12V, but in discrete pulses. If 20% of the time is filled by pulses (20% duty cycle), then the motor will receive small kicks that keep it running at a low percentage of full speed. The motor runs smoothly because of the inertia of the rotor and because the frequency of the pulses can be adjusted.

A PWM is, in some sense, a special-purpose digital-to-analog converter. Most micro-controllers provide one or more PWM output lines. Pulse Width Modulation is also known as Pulse Duration Modulation (PDM).

Three types of Pulse Width Modulation (PWM) are possible:

1. The pulse center may be fixed in the center of the time window and both edges of the pulse moved to compress or expand the width.

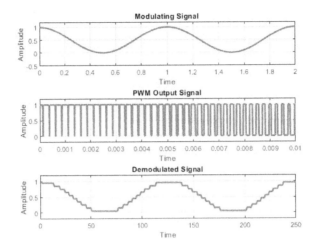

Figure 4.13 The Simulated PWM Output Waveform

2. The lead edge can be held at the lead edge of the window and the tail edge modulated.
3. The tail edge can be fixed and the lead edge modulated.

4.10.2.1 Simulation of PWM in MATLAB

The MATLAB script file to simulate the generation of PWM using the modulate(.) function is given below. The output waveform is shown in Figure 4.13.

```
%% MATLAB script to simulate PWM...
clear all; close all; clf;
fc=4e3; % carrier frequency.
fs=8e4;%sampling frequency..
fm=1e3; % modulating frequency.
t1=linspace(0,8,fm);
x=(1+cos(2*pi*fm*t1))/2; %modulating signal..
[y,t]=modulate(x,fc,fs,'pwm','centered');
x1=demod(y,fc,fs,'pwm','centered');
subplot(311), plot(t1(1:400),x(1:400),LineWidth=2);
axis([0 2 -0.5 1.2]); grid;
xlabel('Time'); ylabel('Amplitude');
title('Modulating Signal');
subplot(312), plot(t,y,LineWidth=2);
axis([0 .01 -0.2 1.2]); grid;
xlabel('Time'); ylabel('Amplitude');
```

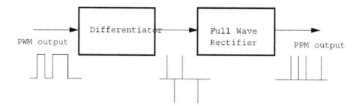

Figure 4.14 Generation of PPM from PWM

```
title('PWM Output Signal');
subplot(313), plot(x1(1:250),LineWidth=2);
axis([0 250 -0.2 1.2]); grid;
xlabel('Time'); ylabel('Amplitude');
title('Demodulated Signal');
```

4.10.2.2 Demodulation of PWM

The demodulation of PWM waveform is achieved by passing it through an integrator circuit. Note that integrator output in this case will be proportional to the width of the pulses. Thus, the PWM waveform is converted to PAM waveform. Demodulation of PAM waveform is done by passing it through a low pass filter, as discussed before. The MATLAB function demod(.) can simulate the demodulation of PWM waveforms.

4.10.3 PULSE POSITION MODULATION (PPM)

The amplitude and width of the pulse are kept constant in the Pulse Position Modulation system. On the other hand, the position of each pulse, in relation to the position of a recurrent reference pulse, is varied by each instantaneous sampled value of the modulating wave. PPM has the advantage of requiring constant transmitter power since the pulses are of constant amplitude and duration. It is widely used but has the disadvantage of depending on transmitter–receiver synchronization.

4.10.3.1 Generation of PPM

PPM can be generated from PWM. As illustrated in figure 4.14, the PWM output is passed through a differentiator and a full wave rectifier to obtain PPM.

4.10.3.2 Simulation of PPM in MATLAB

The MATLAB script file to simulate the generation of PPM using the modulate(.) function is given below. The output waveform is shown in figure 4.15.

```
%% MATLAB script to simulate PPM...
clear all; close all; clf;
```

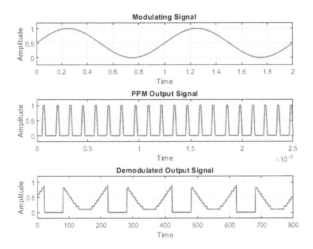

Figure 4.15   The Simulated PPM Output Waveform

```
fc=8e3; % carrier frequency.
fs=8e4;%sampling frequency..
fm=800; % modulating frequency.
t1=linspace(0,4,800);
x=(1+sin(2*pi*fm*t1))/2; %modulating signal..
[y,t]=modulate(x,fc,fs,'ppm',0.2);
x1=demod(y,fc,fs,'ppm');
subplot(311), plot(t1,x,LineWidth=1.5);
axis([0 2 -0.25 1.2]); grid;
xlabel('Time'); ylabel('Amplitude');
title('Modulating Signal');
subplot(312), plot(t,y,LineWidth=1.5);
axis([0 2.5e-3 -0.2 1.2]); grid;
xlabel('Time'); ylabel('Amplitude');
title('PPM Output Signal');
subplot(313), plot(x1,LineWidth=1.5);
axis([0 800 -0.2 1.2]); grid;
xlabel('Time'); ylabel('Amplitude');
title('Demodulated Output Signal');
```

4.10.3.3   Demodulation of Pulse Position Modulation

The Pulse Position Modulation waveform can be demodulated by first converting it back to Pulse Width Modulation and then demodulating the PWM output. The PPM

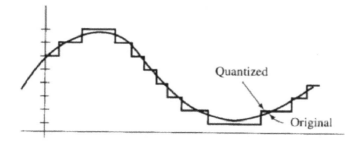

Figure 4.16   The Signal Format of Pulse Code Modulation

can be converted to PWM by triggering a mono-stable multivibrator using the PPM pulses. The PWM can be demodulated by passing it through an integrator.

4.11   PULSE CODE MODULATION (PCM)

The Pulse Code Modulation (PCM) is a digital modulation scheme for transmitting analog data. The signals in PCM are binary; that is, there are only two possible states, represented by logic 1 (high) and logic 0 (low). This is true, no matter how complex the analog waveform happens to be. Using PCM, it is possible to digitize all forms of analog data, including full motion video, voice, music, telemetry, and virtual reality (VR).

The Pulse Code Modulation (PCM) is a direct application of the analog-to-digital converter. Assume that the amplitude of each pulse in a PAM system is rounded off to one of several possible levels. This yields a stair case type of waveform as shown in figure 4.16. The *rounding off* process is known as *quantization* and it introduces an error known as *quantization noise*. The sample values are now transmitted as a binary value.

4.11.1   SAMPLING

In Pulse Code Modulation transmission, before a signal can be quantized it must first be sampled. Sampling is the process of instantaneously capturing the level of a continuous-time (analog) signal at some predetermined rate. This predetermined rate is called the *sampling frequency*. As the sampling rate increases, the sampled signal begins to approximate the original continuous signal. As the sampling frequency decreases, samples move further apart in time, eventually the original signal cannot be reconstructed from the sampled version. The limit on how far apart these samples can be, without losing information, is the basis for *Shannon's Sampling Theorem*. Shannon's sampling theorem was originally stated as follows: "If a function $f(t)$ contains no frequencies higher than f cycles per second, it is completely determined by giving its ordinates at a series of points spaced $(1/2f)$ seconds apart". A mathematical version of this theorem can be obtained by convolving the Fourier transform of the signal to be sampled with the Fourier transform of an infinite sequence of impulses.

$$F_S(\omega) = \frac{1}{2\pi}[F(\omega) * S(\omega)]$$

$$= \frac{1}{2\pi} \int_{-\infty}^{\infty} S(t)F(t-\omega)dt \tag{4.11.44}$$

Where, $F_S(\omega)$ = sampled signal spectrum, $F(\omega)$= original signal spectrum, $S(\omega)$= spectrum for a sequence of impulses, and $\omega = 2\pi f$ (radian frequency).

The result of the convolution in equation 4.11.44 is the original signal spectrum repeated at multiples of the sampling frequency. Notice that if the sampling frequency is less than twice the bandwidth of the original signal, the replicas centered at multiples of the sampling frequency will overlap and distort the original.

This undesirable phenomenon is termed as *aliasing*. To avoid aliasing, it should be maintained that:

$$f_s > 2B.$$

Where, f_s = sampling frequency in Hertz, and B= bandwidth of the original signal in Hertz.

Although the Shannon's sampling theorem is mathematically accurate [2], in most cases it is not practical to sample a signal at exactly twice its highest frequency. Band-limiting is necessary to avoid aliasing. For $f_s = 2B$ the band-limiting filter must have a so-called "brick wall" roll-off at frequency B. A filter that matches this requirement is physically unrealizable. Several factors contribute to the actual sampling frequency used. Generally, there is a compromise between the complexity of the band-limiting filter versus the cost of the analog-to-digital converter (ADC). As f_s becomes closer to $2B$, the band-limiting filter requires more stages to give the desired roll-off. As f_s increases, the required conversion time necessitates a faster ADC. Cost for ADCs is inversely proportional to conversion time; and as conversion time decreases, cost increases.

4.11.1.1   Linear or Uniform Quantization

Figure 4.17 illustrates the transfer characteristics of a seven-level uniform quantizer. To represent the quantized output $q(n)$ as a binary number would require 3 bits. Once a signal has been sampled it is discrete in time. However, the amplitude remains continuous. The quantized version of a signal is obtained by applying the sampled signal to a binary encoder where the discrete voltage levels are assigned to the nearest binary numbers. The complete process, combining sampling and amplitude quantization and encoding, is known as PCM.

---

[2]Consider sampling a sine wave of average value zero and frequency f_0 at $2f_0$ with the sampling instants coinciding with the zero–crossing instants of the sine wave. Then, the result of sampling is a sequence of zeros. This is a *triviality*. So, to avoid this, we make $f_s > 2B$, rather than $f_s \geq 2B$.

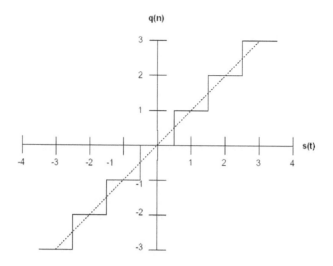

Figure 4.17 Linear Quantizer Transfer Characteristics

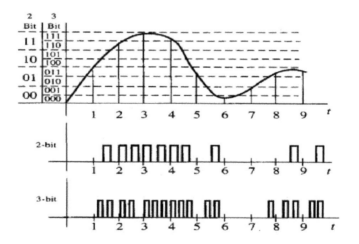

Figure 4.18 The Linear PCM Quantization Process

In PCM, each analog sample is assigned a binary code. Typical binary codes for a 2–bit and 3–bit PCM systems are illustrated in figure 4.18. The digital signal consists of block of n-bits, where each n-bit number is the amplitude of a PCM pulse. Note that the Signal-to-Noise Ratio for quantization noise is given by:

$$SNR_{dB} = 20\log_{10} 2^n + 4.77 dB = 6.02n + 4.77 dB \qquad (4.11.45)$$

Thus, each additional bit increases SNR by 6 dB, or a factor of 4.

The equation 4.11.45 is an objective measure of quality in systems employing uniform quantization. It should be noted that:

- the SNR calculated in equation 4.11.45 is a theoretical maximum. In practice, other factors (e.g. power supply noise) tend to reduce the final SNR.
- Also, the human voice is considered to have about $40dB$ of dynamic range, however, during most conversations it is typically about $20dB$ down from maximum. In other words, we generally do not shout during normal conversation. Consequently, the average signal-to-noise ratio for uniformly quantized speech is about $20dB$ less than what would be calculated using the above equation.

4.11.1.2 Dynamic Range

Another objective measure of quality which can be derived in a similar manner is the Dynamic Range (DR). Dynamic range pertains to the resolution of a quantization scheme. It is the ratio of the full-scale amplitude to the smallest quantized amplitude change:

$$DR = \frac{1}{2}\left(\frac{\alpha 2^n}{\alpha}\right) = 2^{n-1} \tag{4.11.46}$$

The Dynamic Range in dBs is given by:

$$DR_{dB} = 20\log 2^{n-1} = 6.02(n-1). \tag{4.11.47}$$

4.11.1.3 Non-Linear or Non-Uniform Quantization

An alternative to linear quantization is to make the finest resolution α fine for low-level signals and coarse for high-level signals, introducing a non-uniform quantizer characteristic. This type of quantization can result in improved dynamic range for a given number of bits and effectively raise the SNR for lower-level signals. The drawback of non-linear quantization process is a lower value for maximum SNR that can be achieved. The equation 4.11.45 is not valid for non-uniform quantization. Figure 4.19 shows a transfer characteristic for a seven-level non-uniform quantizer. Note that, when compared to the uniform quantizer transfer characteristic shown in figure 4.17, there are three quantization levels for the input $s(t)$ between zero and one, where there is only one with the uniform quantizer.

4.11.2 COMPANDING

A complete digital communication system, incorporating non-uniform speech coding, compresses the quantized signal at the transmitter and then expands the signal back to linear form at the receiver. The compressed signal requires fewer bits and therefore consumes less bandwidth. Ideally, the expander transfer function is the exact inverse of the compressor and is able to reproduce the original uncoded analog signal. In literature pertaining to telecommunications, the words compressor and expander are generally combined into a single word; *compandor*.

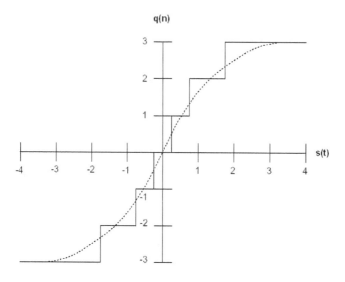

Figure 4.19 Non-Linear Quantizer Transfer Characteristics

Companding is one of the most popular forms of non-uniform quantization. When compared to uniform quantization, it allows for bandwidth compression without degradation in dynamic range, at the expense of peak signal-to-noise ratio. For example, a signal uniformly quantized to 14 bits, would have a peak SNR and DR of approximately $89dB$ and $78dB$ respectively. If the same signal is non-uniformly quantized (compressed) using only 8 bits (256 levels), the minimum quantization level can be set such that DR of $78dB$ can be retained, although the peak SNR will be degraded. Degradation in the peak SNR is due to the course quantization levels used for the large amplitude signals (where fine quantization is not necessary). Low amplitude signals are quantized at finer resolution (more steps, where it is necessary). Consequently, the SNR at low signal levels is improved when compared to uniform quantization with the same number of levels. Essentially, companding strives to make SNR constant over the dynamic range of the quantizer.

4.11.2.1 μ-Law Companding

Most forms of non-uniform quantization are derived from a logarithmic transfer function. That is, the output signal is proportional to the *natural logarithm* of the input signal. In North America and Japan, digital telecommunications networks employ μ-law companding as the standard for PCM encoding ($\mu = 255$ for North America). The general form of the transfer characteristic is:

$$y(x) = sign(x)\frac{ln(1+\mu|x|)}{ln(1+\mu)} \qquad (4.11.48)$$

Where, $-1 \leq x \leq 1$, $x =$ input signal, $y(x) =$ compressed output signal, and $sign(x) =$ polarity of the input signal.

4.11.2.2 A-Law Companding

The A-Law is a standard companding algorithm, used in European digital communications systems to optimize, i.e., modify, the dynamic range of an analog signal for digitizing. It is similar to the μ-Law algorithm.

For a given input x, the equation for A-law encoding is as follows:

$$y(x) \quad = \quad sign(x)\frac{A|x|}{1+\ln(A)}, \quad |x| < \frac{1}{A} \tag{4.11.49}$$

$$= \quad sign(x)\frac{1+\ln(A|x|)}{1+\ln(A)}, \quad \frac{1}{A} \leq |x| \leq 1. \tag{4.11.50}$$

where A is the compression parameter. In Europe, $A = 87.7$; the value 87.6 is also used. The A-law encoding effectively reduces the dynamic range of the signal, thereby increasing the coding efficiency and resulting in a signal-to-distortion ratio that is superior to that obtained by linear encoding for a given number of bits.

The A-law algorithm provides a slightly larger dynamic range than the μ-law at the cost of worse proportional distortion for small signals. By convention, A-law is used for an international connection if at least one country uses it. The A-law is widely used in Europe and the rest of the world, and in India.

4.11.3   DEMODULATION OF PULSE CODE MODULATION

At the destination (receiver end) of the communications circuit, a pulse code demodulator converts the binary numbers back into pulses having the same quantum levels as those in the modulator. These pulses are further processed to restore the original analog waveform.

4.11.4   DIFFERENTIAL PULSE CODE MODULATION (DPCM)

The Differential Pulse Code Modulation (DPCM) is a procedure of converting an analog signal into a digital signal in which an analog signal is sampled and then the difference between the actual sample value and its predicted value (predicted value is based on previous sample or samples) is quantized and then encoded forming a digital value.

The DPCM code words represent differences between samples unlike PCM where code words represented a sample value.

The basic concept of DPCM - coding a difference, is based on the fact that most source signals show significant correlation between successive samples so *encoding uses redundancy in sample values which implies lower bit rate*. The DPCM encoder and decoder are illustrated in figure 4.20. The realization of basic concept described above is based on a technique in which we have to predict current sample value

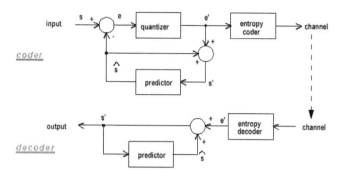

Figure 4.20   The DPCM Encoder and Decoder

based upon previous samples (or sample) and we have to encode the difference between actual value of sample and predicted value (the difference between samples can be interpreted as prediction error). Because it is necessary to predict sample value, DPCM is a form of *predictive coding*.

4.11.4.1   Applications of DPCM

In practice, DPCM is usually used with lossy compression techniques, like coarser quantization of differences can be used, which leads to shorter code words. This is used in JPEG and in adaptive DPCM (ADPCM), a common audio compression method. The ADPCM can be viewed as a superset of DPCM. The good side of the ADPCM method is minimal CPU load, but it has significant quantization noise and only mediocre compression rates can be achieved (4:1).

4.11.4.2   Simulation of DPCM Using MATLAB

The MATLAB functions dpcmenco and dpcmdeco may be used to encode and decode an arbitrary signal in DPCM format. A typical MATLAB script is appended below. The output of the simulation is illustrated in figure 4.21.

```
%% MATLAB Script to simulate DPCM encoder & decoder..
clear all; close all; clf;
predictor = [0 1]; % y(k)=x(k-1)
partition = [-1:.1:.9];
codebook = [-1:.1:1];
t = [0:pi/50:2*pi];
x = sin(4*t); % Original Sinusoidal signal
% Quantize x using DPCM.
encdx = dpcmenco(x,codebook,partition,predictor);
% Try to recover x from the modulated signal.
decdx = dpcmdeco(encdx,codebook,predictor);
plot(t,x,t,decdx,'o',LineWidth=1.5);
```

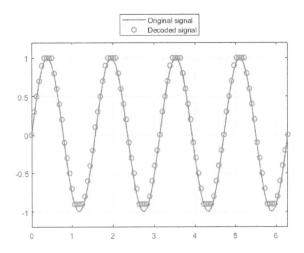

Figure 4.21   Simulation of DPCM Encoder and Decoder

```
grid;axis([0, 2*pi, -1.2 1.2]);
legend('Original signal','Decoded signal','Location','NorthOutside');
mserr = sum((x-decdx).^2)/length(x) % Mean square error
mserr =

    0.0033
```

4.11.5   ADAPTIVE DIFFERENTIAL PULSE CODE MODULATION (ADPCM)

The Adaptive Differential Pulse Code Modulation (ADPCM) is a technique for converting sound or analog information to binary form (a string of 0's and 1's) by taking frequent samples of the sound and expressing the value of the sampled sound modulation in binary terms. ADPCM is used to send sound over fiber optic long-distance cables as well as to store sound along with text, images, and code on a CDROM. Adaptive Differential Pulse Code Modulation was developed for speech coding by P. Cummiskey, Nikil S. Jayant, and James L. Flanagan at Bell Labs in 1973. At bit rates of 24 to 32 kb/s, the ADPCM provides a robust and efficient technique for speech communication and for digital storage of speech. Typical ADPCM encoder and decoder are illustrated in figures 4.22 and 4.23 respectively. In ADPCM quantization step size adapts to the current rate of change in the waveform which is being compressed. Different ADPCM implementations have been explored. The more popular is International Multimedia Association (IMA) ADPCM. The IMA ADPCM implementation is based on the algorithm proposed by International Multimedia Association (IMA). The IMA ADPCM standard specifies compression of PCM from 16 down to 4 bits per sample.

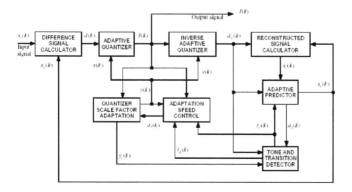

Figure 4.22   The ADPCM Encoder

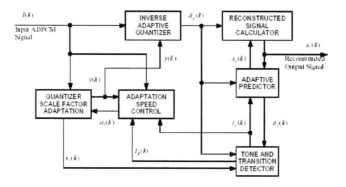

Figure 4.23   The ADPCM Decoder

4.11.5.1   IMA ADPCM

The ADPCM is a lossy compression mechanism. There are various flavors of AD-PCM. The IMA ADPCM algorithm was suggested by the International Multimedia Association (IMA). The IMA ADPCM compresses data recorded at various sampling rates. Sound is encoded as a succession of 4-bit or 3-bit data packets. Each data packet represents the difference between the current sampled signal value and the previous value. The compression ratio obtained is relatively modest. As an example: 16-bit data samples encoded as 4-bit differences result in 4:1 compression format.

The IMA ADPCM is similar to Intel's DVI audio format. The IMA ADPCM is directly supported on most Windows implementations as a native format. Although the quality of IMA ADPCM voice files is not great, the files are portable. There is a real advantage in having compact files that can be played on most Windows PCs. The

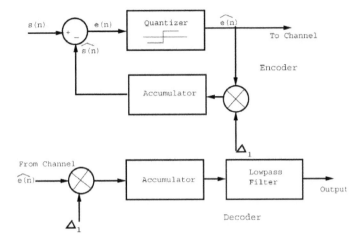

Figure 4.24   The Block Diagram of Delta Modulator Encoder and Decoder

Microsoft Windows supports IMA ADPCM ".wav" files in 4-bit format only and at sample rates between 8 and $41kHz$.

4.11.6   DELTA MODULATION (DM)

The Delta Modulation (DM) is a subclass of Differential Pulse Code Modulation. It can be viewed as a simplified variant of DPCM, in which a 1-bit quantizer is used with the fixed first-order predictor, and was developed for voice telephony applications. The Delta Modulator consists of a 1-bit quantizer, a delay circuit, and two summer circuits, as shown in the block diagram shown in figure 4.24. The delta modulator's output is a stair-case approximated waveform. The delta Δ is the waveform's step size. The waveform's output quality is average.

4.11.6.1   Principle of Delta Modulation

The DM output is 0 if waveform falls in value, 1 represents rise in value, each bit indicates direction in which signal is changing (not how much), i.e. DM codes the direction of differences in signal amplitude instead of the value of difference as in DPCM. The main advantage of Delta Modulation over DPCM is its simple encoder and decoder circuits. This is a direct consequence of a quantizer having only two quantization levels. The basic concept of delta modulation can be explained with the help of the DM encoder and decoder block diagrams as shown in figure 4.24. Typical sinusoidal input waveform, and DM output waveform, when the step size is just right, are illustrated in figure 4.25. Delta Modulation requires "oversampling" in order to obtain an accurate prediction of the next input. Since each encoded sample contains a relatively small amount of information Delta Modulation systems require higher sampling rates than PCM systems. At any given sampling rate, two types

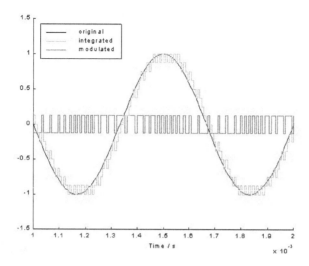

Figure 4.25 The Delta Modulator Input and Output Waveforms

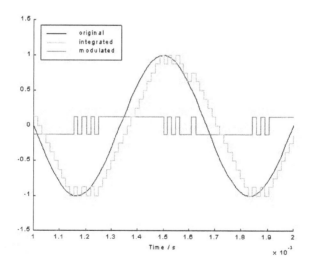

Figure 4.26 Delta Modulation: Illustration of Granular Noise

of distortion, *slope overload distortion* and *granular noise* limit the performance of the DM encoder. The Granular noise and Slope overload distortion are illustrated in figures 4.26 and 4.27 respectively. It may be noted that:

1. Slope overload distortion is due to the use of a step-size delta that is too small to follow portions of the waveform that have a steep slope. It can be reduced by increasing the step size.

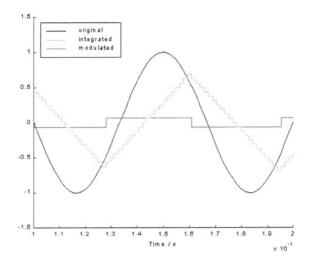

Figure 4.27   Delta Modulation: Illustration of Slope Overload Distortion

2. Granular noise results from using a step size that is too large in parts of the wave-
    form having a small slope. Granular noise can be reduced by decreasing the step
    size.

Delta Modulation is most useful in systems where timely data delivery at the receiver
is more important than the data quality. This modulation is applied to ECG waveform
for database reduction and real-time signal processing.

Even for an optimized step size, the performance of the DM encoder may still be
less satisfactory. An alternative solution is to employ a variable step size that adapts
itself to the short-term characteristics of the source signal. That is the step size is increased when the waveform has a steep slope and decreased when the waveform has
a relatively small slope. This strategy is called Adaptive Delta Modulation (ADM).

4.11.7   ADAPTIVE DELTA MODULATION (ADM)

The Adaptive Delta Modulation (ADM) is similar to Delta Modulation (DM) with
the ability to adjust the slope of the tracking signal, and is widely used in encoding TV and speech signals. The ADM attempts to increase the dynamic range and
the tracking capabilities of fixed step-size DM. In feed-forward adaptation, the step
size of the quantizer is adapted in proportion to the input signal strength. In feedback adaptation, the adaptation is made based on the history of the quantizer as output. Adaptive Delta Modulation method is similar to Delta Modulation except that
the step size is variable according to the input signal in Adaptive Delta Modulation
whereas it is a fixed value in Delta Modulation.

The ADM encoder circuit consists of a summer, quantizer, delay circuit, and a
logic circuit for step size control. The baseband signal $X(nT_s)$ is given as input to the

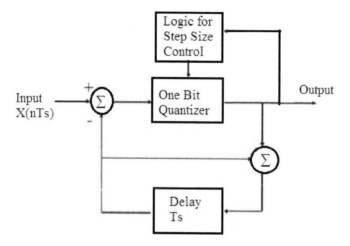

Figure 4.28   The Block Diagram of Adaptive Delta Modulation Encoder

circuit. The feedback circuit present in the transmitter is an Integrator. The integrator generates the staircase approximation of the previous sample. At the summer circuit, the difference between the present sample and staircase approximation of previous sample $e(nT_s)$ is calculated. This error signal is passed to the quantizer, where a quantized value is generated. The step size control block controls the step size of the next approximation based on either the quantized value is high or low. The quantized signal is given as output. Figure 4.28 illustrates the block diagram of the ADM encoder. The ADM decoder has two parts. First part is the step size control. Here the received signal is passed through a logic step size control block, where the step size is produced from each incoming bit. Step size is decided based on present and previous input. In the second part of the receiver, the accumulator circuit recreates the staircase signal. This waveform is then applied to a low pass filter which smoothen the waveform and recreates the original signal.

4.11.7.1   Waveform of Adaptive Delta Modulation

In Adaptive Delta Modulation, the step size of the staircase signal is not fixed and changes depending upon the input signal. Here, first the difference between the present sample value and previous approximation is calculated. This error is quantized; i.e. if the present sample is smaller than the previous approximation, quantized value is *low* or else it is *high*. The output of the one-bit quantizer is given to the *logic step size control* circuit where the step size is decided. At the logic step size control circuit, the output is decided based on the quantizer output. If the quantizer output is high, then the step size is doubled for the next sample. If the quantizer output is low, the step size is reduced by one step for the next sample. A typical ADM output waveform along with the original signal is shown in figure 4.29

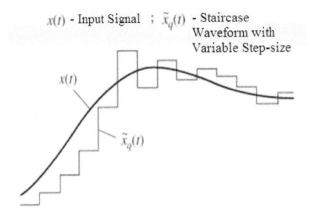

$x(t)$ - Input Signal ; $\tilde{x}_q(t)$ - Staircase
Waveform with
Variable Step-size

$x(t)$

$\tilde{x}_q(t)$

Figure 4.29 The ADM Input and Output Waveforms

4.11.7.2   Advantages of Adaptive Delta Modulation

1. The slope overload error and granular error present in Delta Modulation are solved in Adaptive Delta Modulation. Because of this, the signal-to-noise ratio of ADM is better than DM.
2. During demodulation, ADM uses a low pass filter which removes the quantization noise.
3. In the presence of bit errors, this modulation provides robust performance. This reduces the need for error detection and correction circuits in radio design.
4. The dynamic range of Adaptive Delta Modulation is large as the variable step size covers large range of values.
5. ADM utilizes bandwidth more effectively than Delta Modulation.

4.11.8   SIGMA DELTA MODULATION (SDM)

Sigma Delta Modulator can eliminate the problems associated with the Delta Modulator. The development of Sigma Delta Modulation (SDM) began in the 1960s, to overcome the limitations of delta simulation. Sigma Delta systems quantize the delta (D, difference) between the current sigma and the sigma (sum) of the previous difference (it is a closed-loop feedback system). An integrator is placed at the input of the quantizer; the signal amplitude is constant at a varying frequency (similar to frequency modulation, FM). Thus the SDM is also known as the Pulse Density Modulator (PDM). Similar to Pulse Code Modulation, Sigma Delta Modulation quantizes the signal directly, and not its derivative as in the case of Delta Modulation. Thus the quantization range is dependent upon the maximum signal amplitude and not on the signal spectrum. To achieve high resolution as in the PCM, high sampling rates are

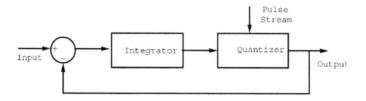

Figure 4.30 The Block Diagram of Sigma Delta Modulator

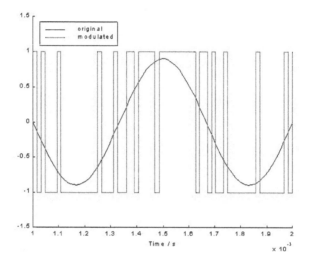

Figure 4.31 The Sigma Delta Modulator Input and Output Waveforms

required. For instance, in CD players with a maximum allowable input signal of $20kHz$ ($44.1kHz$ sampling rate) and $16\times$ oversampling, the internal sampling frequency is at $705kHz$ ($= 44.1kHz \times 16$). In this way, quantization noise is spread from DC ($0Hz$) to $320kHz = 16 \times 20kHz$. SDM adds noise-shaping benefits as the integrator "kills" high-frequency noise.

Figure 4.30 shows the block diagram of a first-order (single integrator) Sigma Delta Modulator encoder. The input to the quantizer is the integral of the difference between the input and the quantized output. When the difference between the input signal and the output signal approaches zero; the average value of the clocked output tracks the input. The integrator forms a lowpass filter on the difference signal thus providing low-frequency feedback around the quantizer. Typical sinusoidal input waveform and Sigma Delta modulated output waveform are illustrated in figure 4.31.

Sigma Delta Modulation achieves high quality by utilizing a negative feedback loop during quantization to the lower bit depth that continuously corrects quantization errors and moves quantization noise to higher frequencies well above the original signal's bandwidth. Subsequent low-pass filtering for demodulation easily removes this high-frequency noise and time averages to achieve high accuracy in amplitude.

Unlike in the case of Pulse Code Modulation and Delta Modulation, the noise is not white, but shaped by a first-order highpass characteristic. In practice, the in-band noise floor level is not satisfactory with first-order SDM. Using a high enough sampling rate low-bit conversion, the in-band noise can be better attenuated (or shaped out) from the audio frequency band that can even exceed the noise immunity that the strict Nyquist rate sampling offers. The latest CDROM audio technology uses Sigma Delta Modulation DAC converters to convert the bits stored in a compact disc to an analog signal that we can hear through a loudspeaker. The CDROM data sampled at $44.1kHz$ is re-quantized and oversampled with a Sigma Delta Modulator and then Continuously Variable Slope Delta modulation back to an analog signal using an SDM decoder (DAC). The advantage of using a $1-bit$ converter in CDROM audio technology is that it offers better tolerance for small variances in the components that is used against the more complex, harder-to-implement multi-bit decoders (and encoders) which use a complex inner structure of voltage references using a network of resistors as described earlier. Since there are so many components, the variance of the value of one component affects the other components (as they form a "chain") thus affecting the overall performance and audio quality reproduction. Another matter of concern is the cost: the $1-bit$ converters offer a better cost-to-performance ratio to multi-bit converters (they can even offer better performance at a lower cost!) that is of great concern in the competing audio and information industry.

4.11.8.1   Applications of Sigma Delta Modulation

Sigma Delta Modulation is the most popular form of analog-to-digital conversion used in several audio applications. It is also commonly used in Digital-to-Analog Converters (DACs), sample-rate converters, and digital power amplifiers. Modern Analog-to-Digital Converters based on Sigma Delta Modulators offer high resolution, high integration, low power consumption, and low cost, making them a good ADC choice for applications such as process control, precision temperature measurements, and weighing scales.

4.11.9   CONTINUOUSLY VARIABLE SLOPE DELTA MODULATION (CVSD)

The Continuously Variable Slope Delta modulation (CVSD) is a differential waveform quantization techniques. It employs two-level quantizers (one bit). Continuously Variable Slope Delta modulation is basically Delta Modulation with an adaptive quantizer. Applying adaptive techniques to a DM quantizer allows for continuous step size adjustment. By adjusting the quantization step size, the coder is able to represent low amplitude signals with greater accuracy (where it is needed) without sacrificing performance on large amplitude signals.

Continuously Variable Slope Delta modulation is used in tactical communications where high communication quality is required; without compromising the

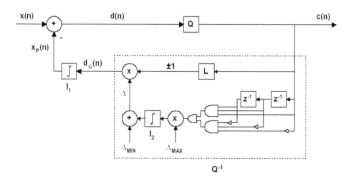

Figure 4.32   The Flow Diagram of CVSD Encoder

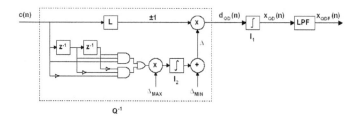

Figure 4.33   The Flow Diagram of CVSD Decoder

security. MIL-STD-188-113 ($16kb/s$ and $32kb/s$), and Federal Standard 1023 ($12kb/s$ CVSD) are examples of a tactical communication systems using Continuously Variable Slope Delta modulation. With the tremendous worldwide growth in wireless technology, secure communication is becoming important to everyone. In addition to point-to-point communication, Continuously Variable Slope Delta modulation is commonly used in digital voice recording or messaging, and audio delay lines. Note that Delta Modulation and Continuously Variable Slope Delta modulation are equivalent to one-bit DPCM and ADPCM, respectively.

Issues with granularity and slope overload can be drastically reduced by making dynamic adjustments to the quantizer step size. Adaptive Delta Modulation (ADM) algorithms attempt to do this by making small Δ for slowly changing signals and large Δ for rapidly changing signals. The most publicized ADM algorithm is the Continuously Variable Slope Delta modulation. It was first proposed by Greefkes and Riemens in 1970. In their CVSD algorithm, adaptive changes in Δ are based on the past three or four sample outputs (i.e. $c(n)$, $c(n-1)$, $c(n-2)$, $c(n-3)$). Figures 4.32 and 4.33 show flow diagrams of the algorithms for the encoder and decoder respectively. Notice only the last three samples of $c(n)$ are used here. The time constants for the integrator $I2$ and integrator $I1$ are typically $4ms$ and $1ms$ respectively. In the literature of Continuously Variable Slope Delta modulation, integrator $I1$ is referred to as the *principal integrator* and integrator $I2$ as the *syllabic integrator*. The so-called syllabic integrator derives its name from the length of *syllable*. Actually, a syllable is nearly $100ms$ in duration, however, pitch changes are on the order of $10ms$. Consequently, $4ms$ seems to work best for the CVSD syllabic (pitch) time constant. The block labeled L performs simple level conversion (i.e. for c(n) = 1, L outputs 1; c(n) = 0, L outputs -1). This CVSD algorithm is also known as "Digitally Controlled Delta Modulation".

4.11.9.1   Applications of Continuously Variable Slope Delta Modulation

CVSD has several attributes that make it well-suited for digital coding of speech. One-bit words eliminate the need for complex framing schemes. Robust performance in the presence of bit errors makes error detection and correction hardware unnecessary. Other speech coding schemes may require a digital signal processing engine and external analog-to-digital/digital-to-analog converters to convert the analog signal into a form that can be processed digitally. The entire CVSD codec algorithm, including input and output filters, can be integrated on a single silicon substrate. Despite this simplicity, CVSD has enough flexibility to allow digital encryption for highly secure applications. Finally, CVSD can operate over a wide range of data rates. It has been successfully used from $9.6kb/s$ to $64kb/s$. At $9.6kb/s$ audio quality is not particularly good, however, it is intelligible. At data rates of $24kb/s$ to $48kb/s$, it is judged as quite acceptable. And above $48kb/s$ it is comparable to toll quality. All of these attributes make CVSD attractive to wireless telecommunication systems (e.g. digital cordless telephones, digital Land Mobile Radio).

The defense industry has been using CVSD for decades in wireline and wireless systems as specified in Mil-Std-188-113. More recently, Federal Standard 1023 proposed CVSD for $25kHz$ channel radios operating above $30MHz$.

4.12   PULSE KEYING TECHNIQUES

Another scheme of digital modulation is using a pulse waveform as the carrier. The original analog baseband signal is converted into a digital baseband signal using an appropriate Analog-to-Digital Converter (ADC). In contrast to analog modulation, where both the baseband (modulating) signal and the carrier are analog, in digital modulation either or both of the carrier and the baseband signals are digital. The digital baseband signal will then modulate the carrier pulse waveform.

4.12.0.1   Comparison of Analog Modulation Schemes with Digital Modulation

In this section, we compare purely analog modulation schemes with digital modulation schemes. The major differences are:

1. Analog modulation schemes must use linear power amplifier stages. This results in reduced efficiencies.
2. Controlling transmitted power and hence the range of communication is difficult.

3. Digital Communication is evolving into more than a necessity in many applications. The baseband is no longer totally analog. The best example is communication involving computer data.
4. By employing suitable data/signal compression and using multiplexing techniques, digital transmission can be made spectrum efficient.

We can modulate the amplitude, frequency, or phase of the carrier pulses. We will consider them in greater detail in the following sections.

4.12.1   AMPLITUDE SHIFT KEYING (ASK)

Amplitude Shift Keying (ASK) is a very popular modulation scheme used in several control applications. This is in part due to its simplicity and low implementation costs. Amplitude Shift Keying modulation has the advantage of allowing the transmitter to idle during the transmission of a "zero", therefore conserving power. The disadvantage of ASK modulation arises in the presence of an undesired signal.

In Amplitude Shift Keying (ASK), the amplitude of the carrier is changed in response to information. In other words, the amplitude of the carrier is switched between two (or more) levels according to the digital data. The information is assumed to be unipolar binary data. Bit 1 is transmitted by a carrier of particular amplitude. To transmit 0, we change the amplitude keeping the frequency constant. The Amplitude Shift Keying waveform can be represented by the following mathematical equation:

$$s(t) = m(t)\sin(2\pi f_c t)$$

where $m(t)$ is the binary message signal, $s(t)$ is the ASK output signal, and f_c is the carrier frequency.

4.12.1.1   Binary ASK or On-Off Keying (OOK)

Binary Amplitude Shift Keying (a.k.a. On-Off Keying) is a special case of Amplitude Shift Keying, where one of the amplitudes is zero as shown in figure 4.34. The MATLAB code used in this simulation is given below:

```
%% MATLAB script to simulate BASK or OOK..
clear all; close all; clf;
fc=8000;
t=linspace(0,1/fc,50);
ec=sin(2*pi*fc*t);
b=mod(randperm(16),2);
n=['The Binary Data is  ', num2str(b)];
ook=[];bin=[];
for i=1:length(b)
    ook=[ook,b(i)*ec];
    bin=[bin,b(i)*ones(1,50)];
end;
```

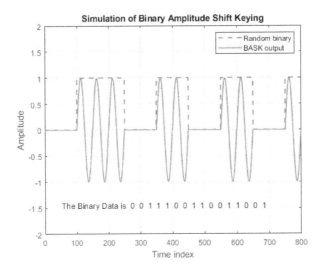

Figure 4.34   Simulation of Binary Amplitude Shift Keying

```
tm=[0:length(ook)-1];
plot(tm,bin,'--',LineWidth=1.5);
axis([0 length(bin) 0 2]); hold on;
plot(tm,ook,LineWidth=1.5);
axis([0 length(tm) -2 2]);hold off;grid;
xlabel('Time index'); ylabel('Amplitude');
legend('Random binary', 'BASK output');
title('Simulation of Binary Amplitude Shift Keying');
gtext(n); % Displays the random binary string..
```

The Amplitude Shift Keying techniques are most susceptible to the effects of non-linear devices which compress and distort the signal amplitude. To avoid such distortion, the system must be operated in the linear range, away from the point of maximum power where most of the non-linear behavior occurs. Despite this problem in high-frequency carrier systems, Amplitude Shift Keying is often used in wire-based radio signaling both with or without a carrier.

4.12.2   FREQUENCY SHIFT KEYING (FSK)

The Frequency Shift Keying (FSK) is the most common form of digital modulation in the high-frequency radio spectrum, and has important applications in telephone circuits. In Frequency Shift Keying (FSK), the instantaneous frequency of the carrier is switched between 2 or more levels according to the baseband digital data. Binary FSK (usually referred to simply as FSK) is a modulation scheme typically used to send digital information between digital equipment such as teleprinters and computers. One frequency is designated as the "mark" frequency and the other as the "space" frequency. The *mark* and *space* frequencies correspond to *binary one* and *zero*, respectively. By convention, mark corresponds to the higher radio frequency.

The minimum duration of a mark or space condition is called the *element length*. Typical values for element length are between 5 and 22 milliseconds, but element lengths of less than 1 microsecond and greater than 1 second have been used. Bandwidth constraints in telephone channels and signal propagation considerations in HF channels generally require the element length to be greater than 0.5 milliseconds. An alternate way of specifying element length is in terms of the keying speed. The keying speed in *bauds* is equal to the inverse of the element length in seconds. For example, an element length of 20 milliseconds (.02 seconds) is equivalent to a 50-baud keying speed.

4.12.2.1   Coherent and Non-coherent FSK

FSK can be transmitted coherently or non-coherently. Coherency implies that the phase of each mark or space tone has a fixed phase relationship with respect to a reference. This is similar to generating an FSK signal by switching between two fixed-frequency oscillators to produce the mark and space frequencies. While this method is sometimes used, the constraint that transitions from mark to space and vice versa must be phase continuous ("glitch" free) requires that the shift and keying rate be interrelated. A synchronous FSK signal which has a shift in Hertz equal to an exact integral multiple ($n = 1, 2, \ldots$) of the keying rate in bauds, is the most common form of coherent FSK. Coherent FSK is capable of superior error performance but non-coherent FSK is simpler to generate and is used for the majority of FSK transmissions. Non-coherent FSK has no special phase relationship between consecutive elements, and, in general, the phase varies randomly.

4.12.2.2   Simulation of BFSK

In the Binary Frequency Shift Keying (BFSK), as mentioned, only two frequencies are involved. The BFSK signal can be represented by

$$
\begin{array}{rcll}
s_0(t) & = & A_c \cos(2\pi f_0 t + \theta_0) \; for \; m(t) = 0 & (4.12.51) \\
s_1(t) & = & A_c \cos(2\pi f_1 t + \theta_1) \; for \; m(t) = 1. & (4.12.52)
\end{array}
$$

where $s_0(t)$ and $s_1(t)$ are the FSK output signals corresponding to binary inputs 0 and 1 respectively. Note that frequencies f_0 and f_1 are chosen for them, respectively. The equations given above will precisely correspond to a discontinuous phase FSK when $\theta_0 \neq \theta_1$. On the other hand, we will get continuous phase FSK when $\theta_0 = \theta_1$. The MATLAB code to simulate Continuous Phase Frequency Shift Keying (CPFSK) is given below. The simulated waveform is illustrated in figure 4.35.

```
%% MATLAB script to Simulate Continuous Phase FSK..
clear all; close all; clf;
f0=2000; % frequency for "0" bits..
```

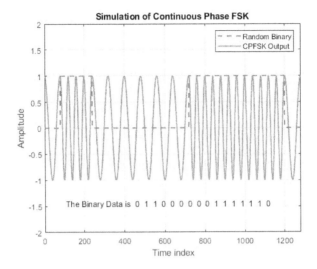

Figure 4.35   Simulated Waveforms for Continuous Phase FSK

```
f1=4000; % frequency for "1" bits..
t=linspace(0,1/f0,80);
e0=cos(2*pi*f0*t); % theta0 =0;
e1=cos(2*pi*f1*t); % theta1 =1;
b=mod(randperm(16),2);
bnot=1-b;
n=['The Binary Data is  ', num2str(b)];
fsk1=[];fsk2=[];bin=[];
for i=1:length(b)
    fsk1=[fsk1,bnot(i)*e0];
    fsk2=[fsk2,b(i)*e1];
    bin=[bin,b(i)*ones(1,80)];
end;
fsk=fsk1+fsk2;
tm=[0:length(fsk1)-1];
plot(tm,bin,'--',LineWidth=1.5);
axis([0 length(bin) 0 2]);hold on;
plot(tm,fsk,LineWidth=1.5);
axis([0 length(tm) -2 2]);hold off;grid;
xlabel('Time index'); ylabel('Amplitude');
legend('Random Binary', 'CPFSK Output');
title('Simulation of Continuous Phase FSK');
gtext(n); % Displays the random binary string..
```

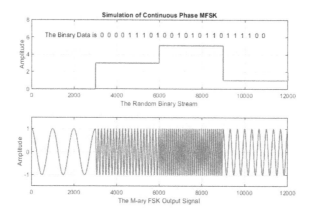

Figure 4.36 Simulated Waveforms for M-ary FSK: (a) The Random Binary Signal. (b) The 8-ary CPMFSK Waveform. The Number of Bits per Symbol in Message = 3; The Total Number of Symbols in Message = 4.

4.12.2.3 M-ary Frequency Shift Keying (MFSK)

M-ary FSK is also known by the name Multiple Frequency Shift Keying. In M-ary Frequency Shift Keying, more than two frequencies are used. M-ary Frequency Shift Keying provides more bandwidth efficiency, but it is more susceptible to error. Mathematically the M-ary Frequency Shift Keying signal can be represented by:

$$
\begin{align}
s_i(t) &= \sqrt{2E/T}\cos(2\pi f_i t),\ \ 1 \leq i \leq M,\ where \tag{4.12.53}\\
f_i &= f_c + (2i - 1 - M)f_d\\
f_c &= the\ carrier\ frequency\\
f_d &= the\ difference\ frequency\\
M &= number\ of\ different\ symbols = 2^L\\
L &= number\ of\ bits\ per\ symbols
\end{align}
$$

To match data rate of input bit stream, each output signal element is held for $T_s = LT$ seconds, where T is the bit period (data rate = $1/T$). Hence, one signal element (symbol) encodes L bits. Note the following points for MFSK:

- Total bandwidth required = $2Mf_d$.
- Minimum frequency separation required = $2f_d = 1/T_s$.
- Therefore, modulator requires a bandwidth of $W_d = 2^L/LT = M/T_s$.

The MATLAB script to simulate Continuous Phase M-ary Frequency Shift Keying (CPMFSK) is given below. The simulated waveform is illustrated in figure 4.36.

```
%% MATLAB script to generate Continuous Phase MFSK..
clear all; close all; clf;
```

```
L=input('The number of bits per symbol in message ');
N=input('The total number of symbols in message ');
M=2^L; % number of different symbols in message stream.
fc=4000; % carrier frequency..
fd=500; % difference frequency...
for i=1:M
    f(i)=fc+(2*i-1-M)*fd;
end;
samp=2*fd;
t=linspace(0,1/f(1),samp);
for j=1:M
    e(j,:)=cos(2*pi*f(j)*t);
end;
b=mod(randperm(2*N*L),2);
n=['The Binary Data is  ', num2str(b)];
mfsk=[];bin=[];
for k=1:L:2*N*L
    l=bin2int(b(k:k+L-1));
    mf=[];bn=[];
    for m=1:L
     mf=[mf, e(l+1,:)];
     bn=[bn,l*ones(1,samp)];
    end;
    mfsk=[mfsk, mf];
    bin=[bin,bn];
end;
tb=[0:length(bin)-1];
subplot(211),plot(tb,bin,LineWidth=1.5);
axis([0 length(tb)*1/2  0 M]);grid;
xlabel('The Random Binary Stream');ylabel('Amplitude');
title('Simulation of Continuous Phase MFSK');
subplot(212),plot(tb,mfsk,LineWidth=1.5);
xlabel('The M-ary FSK Output Signal');ylabel('Amplitude');
axis([0 length(tb)*1/2 -1.5 1.5]); grid;
gtext(n); % Displays the random binary string..
%% bin2int.m function definition
function [y] = bin2int(b)
% function to convert binary numbers to integers.
y=0;
for i=1:length(b)
    y=y+b(i)*2^(length(b)-i);
end;
end
```

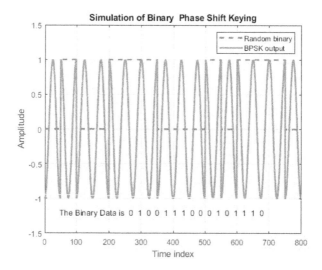

Figure 4.37 Simulated Waveforms for Binary Phase Shift Keying

4.12.2.4   The Essence of Frequency Shift Keying

The key points with regard to Frequency Shift Keying are:

1. Less susceptible to error than ASK.
2. Used for high-frequency (3 to 30 MHz) radio transmission.
3. Can be used at higher frequencies on Local Area Networks (LANs) that use coaxial cable.
4. Amplitude of the carrier wave is constant, and therefore power-efficient.

4.12.3   PHASE SHIFT KEYING (PSK)

In Phase Shift Keying (PSK), the instantaneous phase of the carrier is switched between 2 or more levels according to the baseband digital data. In the Binary Phase Shift Keying (BPSK), which is a subset of Phase Shift Keying, only two phases are involved. The BPSK signal can be represented by

$$s(t) = \begin{cases} \sqrt{2E/T}\cos(2\pi f_c t) & binary \quad 1 \\ \sqrt{2E/T}\cos(2\pi f_c t + \pi) & binary \quad 0 \end{cases} \qquad (4.12.54)$$

$$= \begin{cases} \sqrt{2E/T}\cos(2\pi f_c t) & binary \quad 1 \\ -\sqrt{2E/T}\cos(2\pi f_c t) & binary \quad 0 \end{cases} \qquad (4.12.55)$$

The MATLAB script to simulate Binary Phase Shift Keying (BPSK) is given below. The simulated waveform is illustrated in figure 4.37.

```
%% MATLAB script to simulate Binary PSK..
clear all; close all; clf;
```

```
fc=1000; % frequency for "0" bits..
t=linspace(0,1/fc,50);
e0=cos(2*pi*fc*t);% BPSK output for "1"..
e1=-cos(2*pi*fc*t);% BPSK output for "0".
b=mod(randperm(16),2);
bnot=1-b;
n=['The Binary Data is  ', num2str(b)];
bpsk1=[];bpsk2=[];bin=[];
for i=1:length(b)
    bpsk1=[bpsk1,b(i)*e0];
    bpsk2=[bpsk2,bnot(i)*e1];
    bin=[bin,b(i)*ones(1,50)];
end;
bpsk=bpsk1+bpsk2;
tm=[0:length(bpsk1)-1];
plot(tm,bin,'--',LineWidth=2);
axis([0 length(bin) 0 1.5]);hold on;
plot(tm,bpsk,LineWidth=2);
axis([0 length(tm) -1.5 1.5]);hold off;grid;
xlabel('Time index'); ylabel('Amplitude');
legend('Random binary', 'BPSK output');
title('Simulation of Binary  Phase Shift Keying');
gtext(n); % Displays the random binary string..
```

4.12.4 DIFFERENTIAL PHASE SHIFT KEYING (DPSK)

Differential Phase Shift Keying (DPSK) is regarded as the *noncoherent* version of Binary Phase Shift Keying. It might appear that Differential Phase Shift Keying offers advantages over both ASK, FSK, and PSK. However, the demodulation of these signals is more difficult and hence more expensive. The method of demodulation is an important factor in determining the selection of a modulation scheme. There are two types of demodulation which are distinguished by the need to provide knowledge of the phase of the carrier. Demodulation schemes requiring the carrier phase are termed *coherent*. Those that do not need the phase are termed *incoherent*. Incoherent demodulation can be applied to Amplitude Shift Keying and wide-band Frequency Shift Keying. It describes demodulation schemes that are sensitive only to the power in the signal. With Amplitude Shift Keying, the power is either present, or it is not. With wide band Frequency Shift Keying, the power is either present at one frequency, or the other. Incoherent modulation is inexpensive but has poorer performance. Coherent demodulation requires more complex circuitry, but has better performance. In Amplitude Shift Keying incoherent demodulation, the signal is passed to an envelope detector. This is a device that outputs the "outline" of the signal. A decision is made as to whether the signal is present or not. Envelope detection is the simplest and cheapest method of demodulation.

Incoherent demodulation can be used for wide band Frequency Shift Keying. Here the signals are passed to two circuits, each sensitive to one of the two carrier frequencies. Circuits whose output depends on the frequency of the input are called *discriminators* or *filters*. The outputs of the two discriminators are interrogated to determine the signal. Incoherent Frequency Shift Keying demodulation is simple and cheap, but very wasteful of bandwidth. The signal must be wide band Frequency Shift Keying to ensure that the two signals $f_1(t)$ and $f_2(t)$ are distinguished. It is used in circumstances where bandwidth is not the primary constraint.

The difficulty with coherent detection is the need to keep the phase of the replica signal, termed *local oscillator*, "locked" to the carrier. This is not easy to do. Oscillators are sensitive to (among other things) temperature, and a "free-running" oscillator will gradually drift in frequency and phase.

There are two methods to prevent such an occurrence. In one, a pilot carrier signal is sent in addition to the modulated carrier. This pilot carrier is used to synchronize the local oscillator phase. The alternative is to employ another form of modulation, Differential Phase Shift Keying (DSPK). Differential Phase Shift Keying is actually a simple form of coding. The modulating signal is not the binary code itself, but a code that records changes in the binary code. This way, the demodulator only needs to determine changes in the incoming signal phase. Because the drifts associated with local oscillators are slow, this is not difficult to arrange. The Phase Shift Keying signal is converted to a Differential Phase Shift Keying signal with two rules:

1. a 1 in the Phase Shift Keying signal is denoted by no change in the Differential Phase Shift Keying.
2. a 0 in the Phase Shift Keying signal is denoted by a change in the Differential Phase Shift Keying signal.

The sequence is initialized with a leading 1. An example of the pattern is thus:

$$PSK: \quad 01001101$$
$$DPSK: \quad 100100011$$

The Differential Phase Shift Keying generator can be implemented by using a *delay element* and an *EX-NOR* gate as shown in figure 4.38. The MATLAB code to simulate Differential Phase Shift Keying (DPSK) is given below. The simulated waveform is illustrated in figure 4.39.

```
%% MATLAB/GNU Octave Script file to simulate DPSK generator..
%% It comprises of a pre-processor block and a
%% PSK modulator..
clear all; close all; clf;
fc=2; % carrier frequency;
samp=100;
t=linspace(0,2*pi,samp);
ph1=cos(fc*t);
ph2=-cos(fc*t);
```

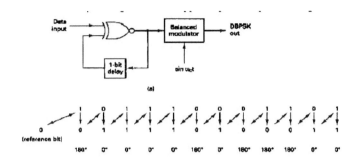

Figure 4.38 The Block Diagram of Differential Phase Shift Keying Generator

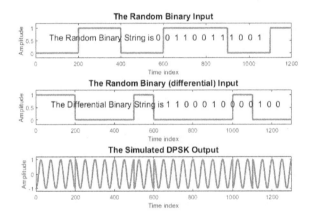

Figure 4.39 Simulated Waveforms for Differential Phase Shift Keying

```
b=mod(randperm(12),2);% The random binary stream..
x=0;
nb(1)=not(xor(b(1),0));
for i=2:length(b)+1
    nb(i)=not(xor(b(i-1),x));
    x=nb(i);
end;
dpsk=[];bin1=[];bin2=[];
for j=1:length(nb)
    if nb(j)==0
        dpsk=[dpsk,ph1];
        bin1=[bin1,zeros(1,samp)];
elseif nb(j)==1
 dpsk=[dpsk,ph2];
```

```
        bin1=[bin1,ones(1,samp)];
    end; % end of if..
end; % end of for

for k=1:length(b)
    if b(k)==0
        bin2=[bin2,zeros(1,samp)];
    elseif b(k)==1
        bin2=[bin2,ones(1,samp)];
    end;
end;
subplot(311),plot(bin2,LineWidth=2);
axis([0 samp*length(b) -0.2 1.2]);
xlabel('Time index'); ylabel('Amplitude');
title('The Random Binary Input','FontSize',12);
bn=num2str(b);
bx=['The Random Binary String is ',bn];
gtext(bx, 'FontSize',12);
subplot(312),plot(bin1,LineWidth=2);
axis([0 samp*length(nb) -0.2 1.2]);hold on;
xlabel('Time index'); ylabel('Amplitude');
title('The Random Binary (differential) Input','FontSize',12);
bn=num2str(nb);
bx=['The Differential Binary String is ',bn];
gtext(bx,'FontSize',12);
subplot(313),plot(dpsk,LineWidth=2);
axis([0 samp*length(nb) -1.2 1.2]);
xlabel('Time index'); ylabel('Amplitude');
title('The Simulated DPSK Output','FontSize',12);
```

4.12.5 M-ARY PHASE SHIFT KEYING (MPSK)

When the number of phases in Phase Shift Keying is more than 2, (i.e. $M > 2$) we have M-ary Phase Shift Keying. We consider the following flavors of M-ary Phase Shift Keying:

1. Quadrature Phase Shift Keying (QPSK).
2. Offset QPSK (Staggered QPSK) (OQPSK/SQPSK).
3. $\pi/4$ Differential QPSK (no carrier) ($\pi/4$ DQPSK).
4. $\pi/4$ Differential QPSK (with carrier) ($\pi/4$ QPSK).
5. Differential MPSK (no carrier recovery) (DMPSK).

4.12.5.1 Quadrature Phase Shift Keying (QPSK)

The Quadrature Phase Shift Keying signal is an extension of the Binary Phase Shift Keying signal. Both of them are special cases of M-ary PSK signals. We can write

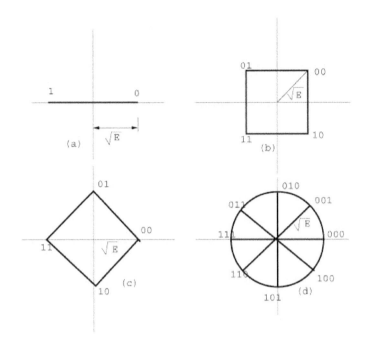

Figure 4.40 The Signal Constellations of Various MPSK Modulations-(a): BPSK (b): QPSK (c): Also QPSK (d): 8-PSK

the process that describes the digital phase modulated signal in a polar form as

$$s_i(t) = A_c \, p_s(t) \cos\left(2\pi f_c t + \frac{2\pi i}{M}\right); i = 0, 1, \ldots M - 1. \qquad (4.12.56)$$

where $p_s(t)$ is the pulse shaping function. In digital phase modulation, the phase of the sinusoid is modified in response to a received bit. A sinusoid can go through a maximum of 2π phase change in one period. So the maximum phase we can change at any one time is 180°. We can use M quantized levels of 2π, to create a variety of PSK modulation. The variable i is a number from $0\, to\, M - 1$. The allowed phases are given by:

$$Modulation\ angle\ \theta_i = \frac{2\pi i}{M}$$

where M stands for the order of the digital phase modulation. $M = 2$ makes it Binary Phase Shift Keying; when $M = 4$, it is Quadrature Phase Shift Keying; when $M = 8$, it becomes $8PSK$, and so on. Figure 4.40 shows three of these digital phase modulations in their "signal constellations". A rotation of the second signal constellation resulting in the third diagram does not change the modulation, its power, or performance. These modulations are called *rotationally invariant*.

For baseband Phase Shift Keying signals, we use a square pulse. The pulse has an amplitude of A. The energy of this pulse is equal to the power of the signal times the

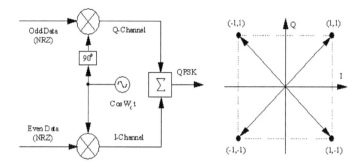

Figure 4.41 The Block Diagram of QPSK Modulator

duration, $T/2$. The power is equal to A^2 with $R = 1\,\Omega$ and $T/2$ is the symbol time (duration). Now, we have $\varepsilon = \frac{A^2 T}{2} = 1$, which is the energy of the pulse. This implies that $A = \sqrt{\frac{2}{T}}$. Thus the pulse has an amplitude of $\sqrt{\frac{2}{T}}$ over a period of $T/2$ seconds. Hence we have:

$$p_s(t) = \sqrt{\frac{2}{T}}, \ 0 \le t \le T \tag{4.12.57}$$

Now, substituting Equation 4.12.57 into Equation 4.12.56, we get:

$$s_i(t) = A_c \sqrt{\frac{2}{T}} \cos\left(2\pi f_c t + \frac{2\pi i}{M}\right) \tag{4.12.58}$$

We now set the carrier amplitude A_c to $\sqrt{E_b}$. Now we have the *modulation equation of a general MPSK signal.*

$$s_i(t) = \underbrace{\sqrt{\frac{2E_b}{T}}}_{Constant} \cos \left(\underbrace{2\pi f_c t}_{\substack{Changing \\ with\ time}} + \underbrace{\frac{2\pi i}{M}}_{\substack{Changing \\ with\ info.}} \right) \quad i = 0, 1, \ldots, M-1. \tag{4.12.59}$$

Now expanding the above equation we can express the QPSK signal as:

$$s(t) = \begin{cases} \sqrt{2E_b/T}\,[\cos(2\pi f_c t)\cos(\pi/4) - \sin(2\pi f_c t)\sin(\pi/4)] & for \quad 11 \\ \sqrt{2E_b/T}\,[\cos(2\pi f_c t)\cos(3\pi/4) - \sin(2\pi f_c t)\sin(3\pi/4)] & for \quad 01 \\ \sqrt{2E_b/T}\,[\cos(2\pi f_c t)\cos(3\pi/4) + \sin(2\pi f_c t)\sin(3\pi/4)] & for \quad 00 \\ \sqrt{2E_b/T}\,[\cos(2\pi f_c t)\cos(\pi/4) + \sin(2\pi f_c t)\sin(\pi/4)] & for \quad 10 \end{cases}$$
$$\tag{4.12.60}$$

A typical QPSK modulator is illustrated in figure 4.41. The cosine term in the equation 4.12.60 is known as the *inphase component* and the sine term is known as the

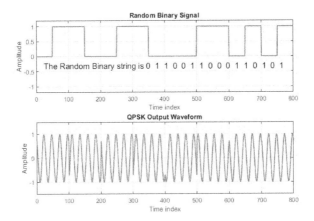

Figure 4.42   Simulated Waveforms for Quadrature Phase Shift Keying

quadrature component of QPSK output. The Quadrature Phase Shift Keying enables efficient utilization of channel bandwidth. In a Quadrature Phase Shift Keying system, note that there are two bits per symbol and the transmitted energy per symbol is twice the signal energy per bit.

4.12.5.2   MATLAB Simulation of QPSK Modulator

The MATLAB script to simulate Quadrature Phase Shift Keying (QPSK) is given below. The simulated waveform is illustrated in figure 4.42.

```
%% MATLAB script file to simulate..
%% QPSK modulation of an arbitrary binary waveform..
clear all; close all; clf;
f=4; % Frequency of binary clock..
sam=100;
t=linspace(0,2*pi,sam);
b=mod(randperm(16),2);% Random binary waveform..
bn=num2str(b);
bx=['The Random Binary string is ',bn];
in=cos(f*t);
qd=sin(f*t);
cp=[];sp=[];
mod1=[];mod2=[];bit=[];
m=sqrt(1/2)*ones(1,sam);
for j=1:2:length(b)
    eb=bin2int(b(j:j+1));
    switch eb
```

```
        case 0,
           m1=-m;
           m2=m1;
           se=zeros(1,sam);
     case 1,
           m1=m;
           m2=-m;
           se=[zeros(1,sam/2),ones(1,sam/2)];
     case 2,
           m1=-m;
           m2=m;
           se=[ones(1,sam/2),zeros(1,sam/2)];
     case 3,
           m1=m;
           m2=m;
           se=ones(1,sam);
        otherwise ;
     end; %% end of switch.
     cp=[cp,m1];
     sp=[sp,m2];
     mod1=[mod1,in];
     mod2=[mod2,qd];
     bit=[bit,se];
end; %end of for..
qpsk=cp.*mod1+sp.*mod2;
subplot(211),plot(bit,LineWidth=1.5);grid;
title('Random Binary Signal');
gtext(bx,'FontSize',12);
axis([0 50*length(b) -1.2 1.2]);
xlabel('Time index');ylabel('Amplitude');
subplot(212),plot(qpsk,LineWidth=1.5);grid;
title('QPSK Output Waveform');
axis([0 50*length(b) -1.5 1.5]);
xlabel('Time index');ylabel('Amplitude');
%% bin2int.m function definition
function [y] = bin2int(b)
% function to convert binary numbers to integers.
y=0;
for i=1:length(b)
    y=y+b(i)*2^(length(b)-i);
end;
end
```

4.12.5.3 Offset Quadrature Phase Shift Keying (Staggered QPSK) (OQPSK/SQPSK)

A variant of Quadrature Phase Shift Keying is Offset Quadrature Shift Keying (OQPSK), or the Staggered Quadrature Phase Shift Keying (SQPSK). As mentioned earlier, the potential for a 180° phase shift in QPSK results in the requirement for better linearity in the power amplifier stages and the potential for spectral re-growth due to the 100% amplitude modulation. Offset Quadrature Phase Shift Keying reduces this tendency by adding a time delay of one-bit period (half a symbol) in the Q-arm (quadrature arm) of the modulator. The result is that the phase of the carrier is potentially modulated every bit (depending on the data), not every other bit as for Quadrature Phase Shift Keying, hence the phase trajectory never approaches the origin. The ability of the modulated signal to demonstrate a phase shift of 180° is therefore removed.

As with the other phase modulation schemes considered, shaping of the phase trajectory between constellation points is typically implemented with a raised cosine filter to improve the spectral efficiency. Due to the similarities between Quadrature Phase Shift Keying and Offset Quadrature Phase Shift Keying, similar signal spectra and probability of error are achieved. The Offset Quadrature Phase Shift Keying modulation is used in the North American IS-95 CDMA cellular telephone system standard for the link from the mobile to the base station.

4.12.6 QUADRATURE AMPLITUDE MODULATION (QAM)

The modulation equation for Quadrature Amplitude Modulation (QAM) is a variant of the one used for Phase Shift Keying. The Quadrature Amplitude Modulation allows changing both the amplitude and the phase. In Phase Shift Keying all points lie on a circle so that the I and Q values are related to each other. The Phase Shift Keying signals are therefore have a constant envelope. If we allow the amplitude to change from symbol to symbol, then we get Quadrature Amplitude Modulation. It can be considered a linear combination of two DSB-SC signals. So it is an AM and PM, simultaneously.

$$s(t) = \underbrace{\sqrt{\frac{2E_s}{T}} \cos\left(\theta(t)\right) \cos\left(2\pi f_c t\right)}_{Inphase\ Component} - \underbrace{\sqrt{\frac{2E_s}{T}} \sin\left(\theta(t)\right) \sin\left(2\pi f_c t\right)}_{Quadrature\ Component} \qquad (4.12.61)$$

The equation 4.12.61 can be used to represent a hybrid modulation type of Quadrature Amplitude Modulation nature. Assume that $M = 16$, so that we have 16 symbols, each representing a 4-bit word. This results in 16-QAM. In QAM, the signal points lie in a rectangle, instead of a circle.

4.12.6.1 16-Quadrature Amplitude Modulation

The signal constellation of 16-QAM is illustrated in figure 4.43. In 16-state Quadrature Amplitude Modulation (16-QAM), there are four I values and four Q values.

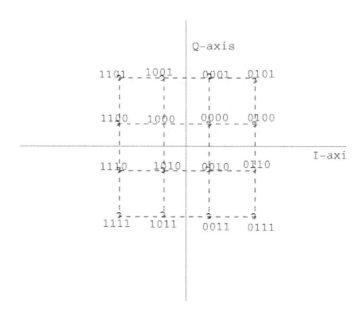

Figure 4.43   The Signal Constellation in the I-Q Plane for 16-QAM

This results in a total of 16 possible states for the signal. It can transition from any state to any other state at every symbol time.

Note that this constellation has 16 points all spread out in the I-Q plane, instead of 16 points all on the circle. It performs better in some situations. Here the points closer to the axes have lesser amplitudes and hence less energy than some others. We can compute the I and Q axis values of each of these points and depending on the total power we want, we can set the value for the amplitude a. For a typical case, we set $a = 1$. The 16-QAM signal mapping is shown in Table 4.2. The symbol rate is one-fourth of the bit rate. So 16-Quadrature Amplitude Modulation format produces a more spectrally efficient transmission. It is more efficient than BPSK, QPSK or 8PSK. Note that QPSK is the same as 4-QAM.

4.12.6.2   64-Quadrature Amplitude Modulation

The 64-Quadrature Amplitude Modulation signal constellation has 64 points all spread out in the I-Q plane. One can prepare a table similar to Table 4.2 for QAM-64 as well. The incoming serial bit stream is converted to a parallel stream, taking 6bits at a time. Note that $2^6 = 64$.

4.12.6.3   256-Quadrature Amplitude Modulation

The 256-Quadrature Amplitude Modulation signal constellation has 256 points all spread out in the I-Q plane. The parallel encoder takes 8bits together. The current

Table 4.2

16-QAM Signal Mapping

Symbol	Bits	Expression	Phase	I	Q
S1	0000	$s_1(t) = \sqrt{\frac{2E_s}{T}}\cos(2\pi f_c t + \pi/4)$	0°	1	1
S2	1000	$s_2(t) = \sqrt{\frac{2E_s}{T}}\cos(2\pi f_c t + 3\pi/4)$	90°	-1	1
S3	1010	$s_3(t) = \sqrt{\frac{2E_s}{T}}\cos(2\pi f_c t + 5\pi/4)$	180°	-1	-1
S4	0010	$s_4(t) = \sqrt{\frac{2E_s}{T}}\cos(2\pi f_c t - \pi/4)$	-90°	1	-1
S5	0100	$s_5(t) = \sqrt{\frac{20E_s}{T}}\cos(2\pi f_c t + 0.10242\pi)$	-26.565^o	3	1
S6	1100	$s_6(t) = \sqrt{\frac{20E_s}{T}}\cos(2\pi f_c t + 0.85242\pi)$	116.56^o	-3	1
S7	1110	$s_7(t) = \sqrt{\frac{20E_s}{T}}\cos(2\pi f_c t + 1.10242\pi)$	153.43^o	-3	-1
S8	0110	$s_8(t) = \sqrt{\frac{20E_s}{T}}\cos(2\pi f_c t - 0.10242\pi)$	-63.435^o	3	-1
S9	0001	$s_9(t) = \sqrt{\frac{20E_s}{T}}\cos(2\pi f_c t + 0.39758\pi)$	26.565^o	1	3
S10	1001	$s_{10}(t) = \sqrt{\frac{20E_s}{T}}\cos(2\pi f_c t + 0.60242\pi)$	63.435^o	-1	3
S11	1011	$s_{11}(t) = \sqrt{\frac{20E_s}{T}}\cos(2\pi f_c t + 1.3976\pi)$	206.57^o	-1	-3
S12	0011	$s_{12}(t) = \sqrt{\frac{20E_s}{T}}\cos(2\pi f_c t + 1.60242\pi)$	243.43^o	1	-3
S13	0101	$s_{13}(t) = \sqrt{\frac{36E_s}{T}}\cos(2\pi f_c t + \pi/4)$	0°	3	3
S14	1101	$s_{14}(t) = \sqrt{\frac{36E_s}{T}}\cos(2\pi f_c t + 3\pi/4)$	90°	-3	3
S15	1111	$s_{15}(t) = \sqrt{\frac{36E_s}{T}}\cos(2\pi f_c t + 5\pi/4)$	180°	-3	-3
S16	0111	$s_{16}(t) = \sqrt{\frac{36E_s}{T}}\cos(2\pi f_c t - \pi/4)$	-90°	3	-3

practical limits are approximately 256-QAM, though work is underway to extend the limits to 512 or 1024 QAM. A 256-QAM system uses 16 I-values and 16 Q-values giving 256 possible states. Since $2^8 = 256$, each symbol can represent eight bits. A 256-QAM signal that can send eight bits per symbol is very spectrally efficient. However, the symbols are very close together and are thus more subject to errors due to noise and distortion. Such a signal may have to be transmitted with extra power (to effectively spread the symbols out more) and this reduces power efficiency as compared to simpler schemes.

4.12.6.4 1024-Quadrature Amplitude Modulation

The 1024-Quadrature Amplitude Modulation signal constellation has 1024 points all spread out in the I-Q plane. The parallel encoder takes 10-bits together. Low redundancy Forward Error Correction (FEC) coded 1024-QAM modems, staggered 1024-QAM, and 256-QAM modems for spectrally efficient (up to 8.84 bits per second per Hertz) microwave and cable systems applications are reported.

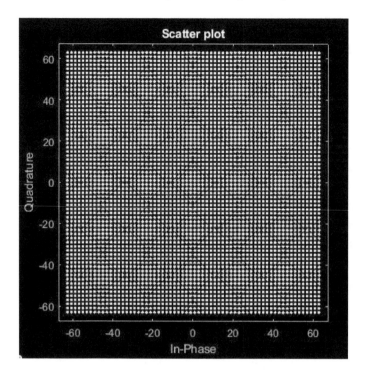

Figure 4.44   The Signal Constellation in the I-Q Plane for 4096-QAM

4.12.6.5   4096-Quadrature Amplitude Modulation

Following are the characteristics of 4096-QAM modulation. It has 12-bits/symbol and a symbol rate of 1/12th of bit rate. It has an increase in capacity of 100% from 64-QAM and 9.77% from 2048-QAM. The number of constellation points in one quadrant is 1024 for 4096-QAM. The constellation diagram of 4096-QAM is illustrated in figure 4.44

4.12.6.6   Advantages of Quadrature Amplitude Modulation

The QAM helps to achieve high data rate as more number of bits are carried by one carrier. Due to this, it has become popular in modern wireless communication system such as LTE, LTE-Advanced etc. It is also used in latest Wireless Local Area Network (WLAN) technologies such as 802.11n 802.11 ac, 802.11ad and others.

4.12.6.7   Disadvantages of Quadrature Amplitude Modulation

- Though data rate has been increased by mapping more than 1 bit on single carrier, it requires high SNR in order to decode the bits at the receiver.
- Needs high linear Power Amplifiers in the Transmitter side.

- In addition to high SNR, higher modulation techniques need very robust front-end algorithms (time, frequency, and channel) to decode the symbols without errors.

4.12.6.8 Simulation of Quadrature Amplitude Modulation in MATLAB

The modulate(.) and qammod(.) functions can be used to simulate QAM. Likewise, the demod(.) and qamdemod(.) functions can be used to demodulate QAM signals. Typical MATLAB script is appended below.

```
%% MATLAB script to simulate QAM using modulate(.) function.
close all; clear all; clc;
fc=80; % carrier frequency.
fs = 200; % sampling frequency
t = 0:1/fs:1;
i = sin(2*pi*10*t) + randn(size(t))/10;
% In-phase component of modulating signal.
q = sin(2*pi*20*t) + randn(size(t))/10;
% Quadrature component of modulating signal.
y = modulate(i,fc,fs,'qam',q); % QAM Generation
pwelch([i;q;y]',hamming(100),80,1024,fs,'centered');
% Plot the Welch Power Spectral Density
legend('In-phase signal','Quadrature signal','QAM Signal');
```

The output of simulation is illustrated in figure 4.45 The application of qammod(.) is illustrated in the following MATLAB script. The scatter plot of the 4096-QAM waveform is shown in figure 4.44.

```
%% MATLAB Script to plot the Signal Constellation
%% of 4096-QAM.
clear all; close all; clf;
M=4096; %
x=[0:M-1];% Modulating signal.
y=qammod(x,M);
scatterplot(y);
```

4.12.7 MINIMUM SHIFT KEYING (MSK)

Coherent detection of modulated signals is also necessary for the perfect demodulation of narrow band Frequency Shift Keying signals. We have already noted that the form of the modulation signal gives rise to a bandwidth that is larger than that required by the Nyquist limit, and that has appreciable energy outside this bandwidth. This is in part due to the abrupt changes in the phase of the signal in Frequency Shift Keying and Phase Shift Keying. By combining coherent detection with appropriate smoothing of the phase change, narrow band Frequency Shift Keying can become

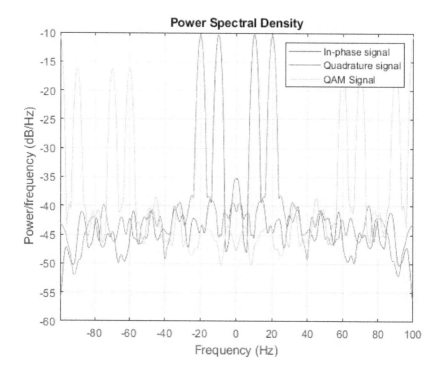

Figure 4.45   The Power Spectral Density of QAM Signal

very bandwidth efficient, approaching the Nyquist limit, and have low side-lobe levels. This form of modulation, which is a combination of Frequency Shift Keying and Phase Shift Keying, is called *Minimum Shift Keying (MSK)*, or *Fast FSK (FFSK)*.

The Minimum Shift Keying is a Continuous Phase Modulation (CPM) scheme where the modulated carrier contains no phase discontinuities and frequency changes occur at the carrier zero crossings. MSK is unique due to the relationship between the frequency of a logical zero and one: the difference between the frequency of a logical zero and a logical one is always equal to half the data rate. In other words, the modulation index is 0.5 for Minimum Shift Keying, and is defined as:

$$m_{msk} = \Delta f \times T$$

where, $\Delta f = |f_{logic1} - f_{logic0}|$, and $T = 1/bitrate$. For example, a 1200-bit per second baseband Minimum Shift Keying data signal could be composed of $1200Hz$ and $1800Hz$ frequencies for a logical one and zero respectively. Baseband Minimum Shift Keying is a robust means of transmitting data in wireless systems where the data rate is relatively low compared to the channel bandwidth.

An alternative method for generating Minimum Shift Keying modulation can be realized by directly injecting Non-Return to Zero (NRZ) data into a frequency modulator with its modulation index set for 0.5. See figure 4.46. This approach is

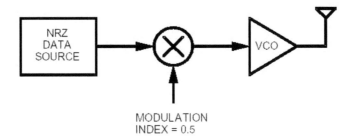

Figure 4.46 The Block Diagram of Direct MSK Generator

essentially equivalent to baseband Minimum Shift Keying. However, in the direct approach the Voltage Controlled Oscillator (VCO) is part of the RF/IF section, whereas in baseband Minimum Shift Keying the voltage to frequency conversion takes place at baseband.

4.12.8 SIMULATION OF MINIMUM SHIFT KEYING WAVEFORMS IN MATLAB

Although MSK is often classified as FM modulation, it is also related to offset-QPSK owing to the dual nature of FSK and PSK modulations. OQPSK is derived from the QPSK by delaying the Q channel by half symbol from I channel. This delay reduces the phase shifts the signal goes through at any one time and results in an *amplifier-friendly* signal.

MSK can be derived from OQPSK by making one further change: The I and Q channels of OQPSK use square root raised cosine pulses. For MSK, change the pulse shape to a half-cycle sinusoid. Figure 4.47 shows an MSK pulse signal and then multiplication by the carrier. In the figure, the carrier signal, the Minimum Shift Keying pulse shape, and the multiplied signal of the pulse shape and the carrier are indicated separately. See the legend for details. The MATLAB code to generate figure 4.47 is given below:

```
%% MATLAB Script for  Simulation of MSK pulse..
clc;clear all; close all; clf;
t=linspace(0,1,200);
f=1;
x=cos(2*pi*f*t);
y=sin(2*pi*f/2*t);
z=x.*y;
x1=sin(2*pi*f*t);
y1=-y;
z1=x1.*y1;
subplot(211),plot(t,x,LineWidth=2);hold on;
plot(t,y,'-',LineWidth=1.5);plot(t,z,'k',LineWidth=3);
```

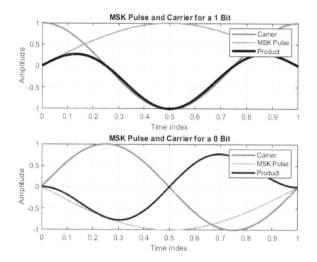

Figure 4.47 (a) MSK Pulse and Carrier for a 1Bit. (b) MSK Pulse and Carrier for 0 Bit

```
xlabel('Time index');ylabel('Amplitude');
title('MSK Pulse and Carrier for a 1 Bit');
grid;legend('Carrier','MSK Pulse','Product');hold off;
subplot(212),plot(t,x1,LineWidth=2);hold on;
plot(t,y1,'-',LineWidth=1); plot(t,z1,'k',LineWidth=2);
xlabel('Time index');ylabel('Amplitude');
title('MSK Pulse and Carrier for a 0 Bit');
grid;legend('Carrier','MSK Pulse','Product');hold of;
%%% end of mskpuls.m
```

The carrier signal expression for Minimum Shift Keying is

$$c(t) = a(t)\sin\left(\frac{\pi}{2T}t\right)\cos\left(\frac{\pi}{T}t\right) + a(t)\sin\left(\frac{\pi}{2T}t\right)\sin\left(\frac{\pi}{T}t\right) \qquad (4.12.62)$$

where the underlined portion gives the half-sinusoid pulse shape. In Minimum Shift Keying, the shape is continuously changing, so there is no discrete jump in the modulated signal at the symbol edge as there is in QPSK. In figure 4.48, the dashed lines are the QPSK I and Q channel symbols of QPSK and the solid line shows how these have been shaped by the half-sine wave. The I and Q channels are computed by

$$MSKI(t) = QPSKI(t)\left|\sin\left(\frac{\pi t}{2T}\right)\right|$$

$$MSKQ(t) = QPSKQ(t)\left|\sin\left(\frac{\pi t}{2T}\right)\right|$$

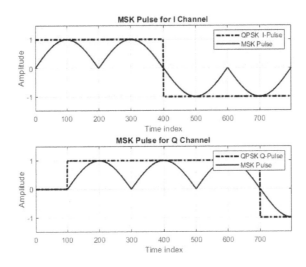

Figure 4.48 (a) MSK Pulse for I Channel. (b) MSK Pulse for Q Channel

The MATLAB code to generate the plot in figure 4.48 is given below:

```
%%% Minimum Shift Keying pulse Carrier simulation..
clc;close all; clear all; clf;
t=linspace(0,2,200);
t1=linspace(0,1,length(t)/2);
f=1;
x=sin(2*pi*f/4*t);%half wave sinusoid..
x1=[x,x,x,x];
y=[zeros(1,length(t)/2),x,x,x,sin(2*pi*f/4*t1)];
n=ones(1,length(t));
i=[n,n,-n,-n];
q=[zeros(1,length(t)/2),n,n,n,-ones(1,length(t)/2)];
ic=x1.*i;
qc=y.*q;
nx=[0:length(i)-1];
subplot(211),plot(nx,i,'k -.',LineWidth=2);hold on;
plot(nx,ic,'k',LineWidth=1,5);hold off;
xlabel('Time index');ylabel('Amplitude');
axis([0 length(i)-1 -1.5 1.5]);grid;
title('MSK Pulse for I Channel');
legend('QPSK  I-Pulse','MSK Pulse');
subplot(212),plot(nx,q,'k -.',LineWidth=2);hold on;
plot(nx,qc,'k',LineWidth=1.5);hold off;
xlabel('Time index');ylabel('Amplitude');
```

```
axis([0 length(i)-1 -1.5 1.5]);grid;
title('MSK Pulse for Q Channel');
legend('QPSK Q-Pulse','MSK Pulse');
%%% end of mskcar.m
```

The I and Q channels are then multiplied by the carrier, cosine for I channel, and sine for the Q channel. Note that the period of the pulse shape is twice that of the symbol rate.

$$MSKcarrI(t) = QPSKI(t)\left|\sin\left(\frac{\pi t}{2T}\right)\right|\cos\left(\frac{\pi t}{T}\right)$$

$$MSKcarrQ(t) = QPSKQ(t)\left|\sin\left(\frac{\pi t}{2T}\right)\right|\sin\left(\frac{\pi t}{T}\right)$$

The Frequency Shift Keying and Minimum Shift Keying produce constant envelope carrier signals, which have no amplitude variations. This is a desirable characteristic for improving the power efficiency of transmitters. Amplitude variations can exercise nonlinearities in an amplifier as amplitude-transfer function, generating spectral regrowth, a component of adjacent channel power. Therefore, more efficient amplifiers (which tend to be less linear) can be used with constant-envelope signals, reducing power consumption.

The MSK has a narrower spectrum than wider deviation forms of FSK. The width of the spectrum is also influenced by the waveforms causing the frequency shift. If those waveforms have fast transitions or a high slew rate, then the spectrum of the transmitter will be broad. In practice, the waveforms are filtered with a Gaussian filter, resulting in a narrow spectrum. In addition, the Gaussian filter has no time-domain overshoot, which would broaden the spectrum by increasing the peak deviation. MSK with a Gaussian filter is termed GMSK (Gaussian MSK).

4.12.8.1 Limitations of Minimum Shift Keying

The fundamental problem with Minimum Shift Keying is that the spectrum is not compact enough to realize data rates approaching the RF channel bandwidth. A plot of the spectrum for Minimum Shift Keying reveals side lobes extending well above the data rate (see figure 4.49). For wireless data transmission systems which require more efficient use of the RF channel bandwidth, it is necessary to reduce the energy of the MSK upper side lobes. Earlier we stated that a straightforward means of reducing this energy is lowpass filtering the data stream prior to presenting it to the modulator (pre-modulation filtering). The pre-modulation lowpass filter must have a narrow bandwidth with a sharp cutoff frequency and very little overshoot in its impulse response. This is where the Gaussian filter characteristic comes in. It has an impulse response characterized by a classical Gaussian function (bell-shaped curve), as shown in figure 4.49. Notice the absence of overshoot or ringing. Figure 4.49 depicts the impulse response of a Gaussian filter for $BT = 0.3$ and 0.5. It may be noted that BT is the product of the bandwidth of the Gaussian filter and time interval between different bits (which is $\frac{1}{Bit\ Rate}$). Thus, the BT is related to the filter's 3dB

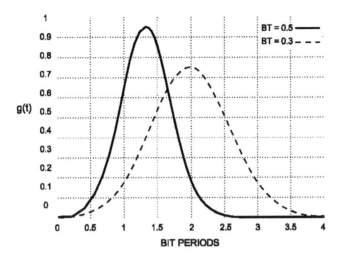

Figure 4.49 The Gaussian Filter Impulse Response for $BT = 0.3$ *and* 0.5. $BT = \frac{f_{3dB}}{Bit\,Rate}$

bandwidth and data rate by

$$BT = \frac{f_{3dB}}{Bit\,Rate} \qquad (4.12.63)$$

Hence for a data rate of 9.6 *kbps* and a BT of 0.5, the filter's 3dB cutoff frequency is 4800 *Hz*.

4.12.9 GAUSSIAN MINIMUM SHIFT KEYING (GMSK)

The proliferation of computers in today's society has increased the demand for transmission of data over wireless links. Binary data, composed of sharp "one to zero" and "zero to one" transitions, results in a spectrum-rich in harmonic content that is not well suited to RF transmission. Hence, the field of digital modulation has been flourishing, in terms of bandwidth efficiency and power requirements. Recent standards such as Cellular Digital Packet Data (CDPD) and Mobitex specify Gaussian filtered Minimum Shift Keying (GMSK) for their modulation method.

The Gaussian Minimum Shift Keying (GMSK) is a simple yet effective approach to digital modulation for wireless data transmission. To provide a good understanding of GMSK, we will review the basics of MSK and GMSK, as well as how GMSK is implemented in CDPD and Mobitex systems. GMSK modems reduce system complexity, and in turn lower system cost. There are, however, some important implementation details to be considered.

4.12.9.1 The Working Principle of GMSK

If we look at a Fourier series expansion of a data signal we see harmonics extending to infinity. When these harmonics are summed, they give the data signal its sharp transitions. Hence, an unfiltered Non-Return to Zero (NRZ) data stream used to modulate an RF carrier will produce an RF spectrum of considerable bandwidth. Of course, the FCC has strict regulations about spectrum usage and such a system is generally considered impractical. But if we remove the high-frequency harmonics from the Fourier series (i.e. pass the data signal through a lowpass filter), the transitions in the data will become progressively less sharp. This suggests that pre-modulation filtering is an effective method for reducing the occupied spectrum for wireless data transmission. In addition to a compact spectrum, a wireless data modulation scheme must have good bit error rate (BER) performance under noisy conditions. Its performance should also be independent of power amplifier linearity to allow the use of class C power amplifiers.

It may be noted that, in general, schemes that rely on more than two levels (e.g. QAM, QPSK) require better signal-to-noise ratios (SNR) than two-level schemes for similar BER performance. Additionally, in a wireless environment, multi-level schemes generally require greater power amplifier linearity than two-level schemes. The fact that GMSK uses a two-level continuous phase modulation (CPM) format has contributed to its popularity. Another point in its favor is that it allows the use of class C power amplifiers (relatively non-linear) and data rates approaching the channel bandwidth (dependent on filter bandwidth and channel spacing).

4.12.9.2   Demodulation of Gaussian Minimum Shift Keying (GMSK)

Demodulation of the GMSK signal requires as much attention to the preservation of an unadulterated wave form as does modulation of the signal. The choice of a Gaussian-shaped pre-modulation filter was made for three main reasons:

1. narrow bandwidth and sharp cutoff.
2. impulse response with low overshoot.
3. preservation of the filter output pulse area.

The first condition gives GMSK modulation its spectral efficiency. It also improves its noise immunity when demodulating. The second condition affords GMSK low phase distortion. This is a major concern when the receiver is demodulating the signal down to baseband, and care must be taken in the design of the IF filtering to protect this characteristic. The third condition ensures the coherence of the signal. While this is quite strict and not realizable with a physical Gaussian filter, the phase response can be kept linear and therefore sufficient for coherent demodulation.

In most systems, the constraints on the above goals also include:

- Data Rate.
- Transmitter-filter bandwidth (BT).
- Channel Spacing.
- Allowable adjacent channel interference.
- Peak carrier deviation.
- Transmitter and Receiver carrier frequency accuracy.

Table 4.3
Theoretical Bandwidth Efficiency Limits

Modulation Format	Theoretical Bandwidth Efficiency
BPSK	1bit/second/Hz
QPSK	2bits/second/Hz
8PSK	3bits/second/Hz
MSK	1bit/second/Hz
16-QAM	4bits/second/Hz
256-QAM	8bits/second/Hz
1024-QAM	10bits/second/Hz
4096-QAM	12bits/second/Hz

- Modulator and Demodulator linearity.
- Receiver IF filter frequency and phase characteristics.

These constraints are all part of the balance that must be struck to provide a robust GMSK system. The data rate, Transmitter BT, peak carrier deviation, and carrier frequency accuracy between receiver and transmitter all contribute to the necessary width of the IF filter. The IF filter should have sufficient width to accommodate the maximum variations in the above parameters so that the received signal will not run into the skirts of the filter. The skirts of the IF filter can introduce excessive amounts of group delay (phase distortion) in the higher frequency components of the received data. The passband of the IF filter should have little or no group delay. The more group delay introduced, the more degraded the bit error rate (BER) performance of the receiver will become. Rules of thumb for group delay dictate less than 10% of a bit time is tolerable. How happy we are with this level of performance is very dependent on the other factors that influence the BER of our system: BT, signal strength, fading, etc. Phase equalization measures can also be taken to help reduce group delay, but if there is control over the IF filter's design these steps can be avoided.

4.12.10 THEORETICAL BANDWIDTH EFFICIENCY LIMITS

The bandwidth efficiency describes how efficiently the allocated bandwidth is utilized or the ability of a modulation scheme to accommodate data, within a limited bandwidth. Table 4.3 shows the theoretical bandwidth efficiency limits for the main modulation types. Note that these figures cannot actually be achieved in practical radios since they require perfect modulators, demodulators, filter, and transmission paths. If the radio had a perfect (rectangular in the frequency domain) filter, then the occupied bandwidth could be made equal to the symbol rate.

Techniques for maximizing spectral efficiency include the following:

1. Relate the data rate to the frequency shift, as in Global System for Mobile (GSM).
2. Use premodulation filtering to reduce the occupied bandwidth.

3. Raised cosine filters, as used in North American Digital Cellular system (NADC), Pacific Digital Cellular system (PDC), and Personal Handyphone System (PHS) give the best spectral efficiency.
4. Restrict the types of transitions.

4.12.10.1 Spectral Efficiencies in Practical Radio Systems

The following examples indicate spectral efficiencies that are achieved in some practical radio systems.

The Time Division Multiple Access (TDMA) version of the North American Digital Cellular (NADC) system, achieves a 48 kbits per second data rate over a 30 kHz bandwidth or 1.6 bits per second per Hz. It is a $\pi/4$ Differential Quadrature Phase Shift Keying-based system and transmits two bits per symbol. The theoretical efficiency would be two bits per second per Hz and in practice it is 1.6 bits per second per Hz.

Another example is a microwave digital radio using $16-QAM$. This kind of signal is more susceptible to noise and distortion than something simpler such as Quadrature Phase Shift Keying. This type of signal is usually sent over a direct line-of-sight microwave link or over a wire where there is very little noise and interference. In this microwave-digital-radio example the bit rate is 140 Mbits per second over a very wide bandwidth of 52.5 MHz. The spectral efficiency is 2.7 bits per second per Hz. To implement this, it takes a very clear line-of-sight transmission path and a precise and optimized high-power transceiver.

4.12.11 PERFORMANCE COMPARISON AND APPLICATIONS

In this section, we compare the performance of various modulation formats discussed so far. We first consider the definition of bit rate and symbol (baud) rate.

4.12.11.1 Bit Rate and Symbol Rate

To understand and compare different modulation format efficiencies, it is important to first understand the difference between bit rate and symbol rate. The bit rate defines the rate at which information is passed. The baud (or signaling) rate defines the number of symbols per second. Each symbol represents n bits, and has M signal states, where $M = 2^n$. This is called M-ary signaling. The signal bandwidth needed for the communications channels depends on the symbol rate, not on the bit rate. The symbol rate is defined as:

$$Symbol\ Rate = \frac{Bit\ Rate}{The\ number\ of\ bits\ transmitted\ with\ each\ symbol} \qquad (4.12.64)$$

Bit rate is the frequency of a system bit stream. Take, for example, a radio with an 8-bit sampler, sampling at 10 kHz for voice. The bit rate, the basic bit stream rate in the radio, would be eight bits multiplied by 10k samples per second, or 80 kbps. (For the moment we will ignore the extra bits required for synchronization, error correction, etc.). The symbol rate is the bit rate divided by the number of bits that can be transmitted with each symbol. If one bit is transmitted per symbol, as with BPSK, then the symbol rate would be the same as the bit rate of 80 kbps. If two bits are transmitted per symbol, as in QPSK, then the symbol rate would be half of the bit rate or 40 kbps. Symbol rate is sometimes called *baud rate*. Note that baud rate is not the same as bit rate. These terms are often confused. If more bits can be sent with each symbol, then the same amount of data can be sent in a narrower spectrum. This is why modulation formats that are more complex and use a higher number of states can send the same information over a narrower piece of the RF spectrum.

4.12.11.2   The Maximum Data Rate

The maximum rate of information transfer through a baseband channel is given by:

$$\textit{Channel Capacity, } C = 2W \log_2 M \textit{ bits per second} \qquad (4.12.65)$$

where W = bandwidth of modulating baseband signal.

4.12.11.3   The Shannon-Hartley Theorem on Channel Capacity

For error-free communication, it is possible to define the capacity which can be supported in an additive white Gaussian noise (AWGN) channel. The channel capacity is given by:

$$C = \log_2(1 + E_b f_b / \eta W) \qquad (4.12.66)$$

where f_b = Capacity (bits per second)
   W = bandwidth of the modulating baseband signal (Hz)
   E_b = energy per bit
   η = noise power density (watts/Hz). Thus, $E_b f_b$ is the total signal power and ηW is the total noise power. The quantity f_b/W is known as the *bandwidth efficiency* (bits per second per Hertz). The Equation 4.12.66 is referred to as the Shannon-Hartley Theorem on channel capacity. A comparison of various modulation formats based on the logarithm of bandwidth efficiency (*to the base 2*), and error-free E_b/N_o ratio is given in Table 4.4.

4.12.11.4   The Symbol Clock

The symbol clock represents the frequency and exact timing of the transmission of the individual symbols. At the symbol clock transitions, the transmitted carrier is at the correct I/Q (or magnitude/phase) value to represent a specific symbol (a specific point in the constellation).

In any digital modulation system, if the input signal is distorted or severely attenuated the receiver will eventually lose symbol lock completely. If the receiver can no longer recover the symbol clock, it cannot demodulate the signal or recover any information. With less degradation, the symbol clock can be recovered, but it is noisy, and the symbol locations themselves are noisy. In some cases, a symbol will fall far

Table 4.4

Comparison of Various Modulation Formats

Modulation Format	$\log_2(Bandwidth\ Efficiency)$	Error Free E_b/N_o
BFSK	0	13dB
BPSK	0	10.6dB
4-QAM	1	10.1dB
4PSK	1	10.1dB
8PSK	1.59	14.5dB
16-QAM	2	15dB
16PSK	2	18dB

Table 4.5

Applications of Various Modulation Formats

Modulation Format	Applications
MSK, GMSK	GSM, CDPD
BPSK	Deep Space Telemetry, Cable Modems
QPSK, $\pi/4$ DQPSK	Satellite, CDMA, NADC, TETRA, PHS, PDC, LMDS, DVB-S, Cable Modems
OQPSK	CDMA, Satellite
FSK, GFSK	DECT, Paging, RAM Mobile Data, AMPS, CT2, Land Mobile
8, 16 VSB	North American DTV, Broadcast, Cable
8PSK	Satellite, Aircraft, Telemetry
16-QAM	Microwave Digital Radio, Modems, DVB-C, DVB-T
32-QAM	Terrestrial Microwave, DVB-T
64-QAM	DVB-C, Modems, MMDS, Digital CATV Systems
256-QAM	Modems, DVB-C (Europe), Digital Video (US)
4096-QAM	802.11 WiFi

enough away from its intended position that it will cross over to an adjacent position. The I and Q level detectors used in the demodulator would misinterpret such a symbol as being in the wrong location, causing bit errors. QPSK is not as efficient, but the states are much farther apart and the system can tolerate a lot more noise before suffering symbol errors. QPSK has no intermediate states between the four corner-symbol locations so there is less opportunity for the demodulator to misinterpret symbols. QPSK requires less transmitter power than QAM to achieve the same bit error rate.

4.12.11.5 Applications of Various Modulation Formats

Table 4.5 covers the applications for different modulation formats in both wireless communications and video.

4.12.12   THE COMPETING GOALS OF SPECTRAL EFFICIENCY AND POWER CONSUMPTION

As with any natural resource, it makes no sense to waste the RF spectrum by using channel bands that are too wide. Therefore narrower filters are used to reduce the occupied bandwidth of the transmission. Narrower filters with sufficient accuracy and repeatability are more difficult to build.

4.12.12.1   Means to Achieve Better Spectral Efficiency

Since spectral efficiency is bits transmitted per second divided by available signal bandwidth, if the overhead could be reduced, the spectral efficiency would go up. But an important part of overhead is the redundant bits added at the channel encoder for Forward Error Correction (FEC), which provides data transmission robustness. The other pieces and parts of transmitted overhead are important, too. For the sake of this discussion, assume that reducing overhead is not practical. Another way to improve spectral efficiency is to use a higher-order modulation. If the channel characteristics can support, say, 1024-QAM (10 bits per symbol), and assuming a symbol rate of 53,60,537symbols/s, the raw data bit-rate is $53,60,537 symbols/s \times 10 bits/symbol = 53,605,370 bps$, or about $53.61 Mbps$. Subtracting the overhead, this scheme yields a net data rate of somewhere around $47.5 Mbps$. In this example, the approximate spectral efficiency is $\frac{47,500,000 bps}{6,000,000 Hz} = 7.92 bits/s/Hz$.

Exercise 4.1. Obtain the spectral efficiency of a 4096-QAM system, when the symbol rate is 53,60,537symbols/s and the redundant bits/s introduced during the transmission as part of Forward Error Correction is 2048000bits/s. Assume a channel bandwidth of $6 MHz$.

Solution. For 4096-QAM, there will be 12bits/symbol; hence gross data rate is $53,60,537 symbols/s \times 12 bits/symbol = 64,326,444 bits/s$. Subtracting the redundant bits/sec used for FEC, the net bits transmitted per second is 62,278,444bits/s. Hence, the *spectral efficiency* is $\frac{62,278,444 bits/s}{6,00,000 MHz} = 103.797 bits/s/Hz$.

4.12.12.2   Relation Between Bit Rate and Baud Rate

Bit rate is the transmission of a number of bits sent per second. On the other hand, the Baud rate is defined as the number of signal units (symbols) sent per second. The formula which relates both bit rate and the baud rate is:

$$Bit\ Rate = Baud\ Rate \times the\ number\ of\ bits\ per\ Baud.$$

In Amplitude Shift Keying (ASK), the bit rate and baud rate are equal.

4.12.13   SQUARE-ROOT-RAISED-COSINE FILTER (SRRC)

In signal processing, a root-raised-cosine filter (RRC), sometimes known as square-root-raised-cosine filter (SRRC), is frequently used as the transmit and receive filter in a digital communication system to perform matched filtering. The combined response of two such filters is that of the raised-cosine filter (RCF). It obtains its name from the fact that its frequency response, $H_{rrc}(f)$, is the square root of the frequency response of the raised-cosine filter, $H_{rc}(f)$:

$$H_{rc}(f) = H_{rrc}(f).H_{rrc}(f) \qquad (4.12.67)$$

The impulse response of the root-raised cosine filter is given by:

$$h(t) = \begin{cases} 1 - \beta + 4\frac{\beta}{\pi}; & t = 0 \\ \frac{\beta}{\sqrt{2}}\left[\left(1 + \frac{2}{\pi}\right)\sin\left(\frac{\pi}{4\beta}\right) + \left(1 - \frac{2}{\pi}\right)\cos\left(\frac{\pi}{4\beta}\right)\right]; & t = \pm\frac{T_s}{4\beta} \\ \frac{\sin\left[\pi\frac{t}{T_s}(1-\beta)\right] + 4\beta\frac{t}{T_s}\cos\left[\pi\frac{t}{T_s}(1+\beta)\right]}{\pi\frac{t}{T_s}\left[1 - \left(4\beta\frac{t}{T_s}\right)^2\right]}; & otherwise. \end{cases} \qquad (4.12.68)$$

It should be noted that unlike the raised-cosine filter, the impulse response is not zero at the intervals of $\pm T_s$. However, the combined transmit and receive filters form a raised-cosine filter which does have zero at the intervals of $\pm T_s$. Only in the case of $\beta = 0$, does the root raised-cosine have zeros at $\pm T_s$.

4.12.14   THE RAISED COSINE FILTER (RCF)

The Raised Cosine Filter (RCF) is characterized by two factors: the roll-off factor, β, and T_s, the reciprocal of the symbol-rate. The impulse response of the raised cosine filter is given by:

$$h(t) = \frac{\sin(\pi t/T_s)}{\pi t} \cdot \frac{\cos(\pi t \beta/T_s)}{1 - 4\beta^2 t^2/T_s^2}. \qquad (4.12.69)$$

where, t is the time, T_s is the symbol rate, and β is the *rolloff factor*. There are several other valid forms of the above equation. A plot of the impulse response of the raised cosine filter is shown in figure 4.50 and the MATLAB code to generate the same is appended below.

```
%%% Raised Cosine filter Impulse Response..
clear all; close all; clf;
t=[-2:.001:2];
Ts=800;
beta=0.8;
h=(sin(pi*t/Ts)./(pi*t)).*(cos(pi*t*beta/Ts)...
    ./(1-4*beta^2*t.^2/Ts^2));
plot(t,h,'Linewidth',3);grid
xlabel('Time, t'); ylabel('Amplitude');
title('The Impulse Response of the Raised Cosine Filter');
%%% end of rascos.m
```

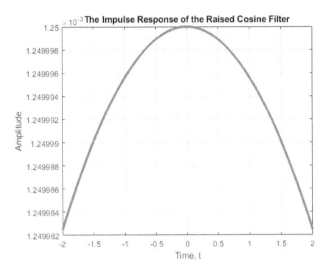

Figure 4.50 The Impulse Response of the Raised Cosine Filter

4.12.14.1 The α Value of the Raised Cosine Filter

The sharpness of a raised cosine filter is described by its alpha (α). Alpha gives a direct measure of the occupied bandwidth of the system and is calculated as:

$$Occupied\ bandwidth = symbol\ rate \times (1+\alpha). \qquad (4.12.70)$$

If the filter had a perfect (brick wall) characteristic with sharp transitions and an alpha of zero, the occupied bandwidth would be equal to symbol rate. Smaller values of alpha increase Inter Symbol Interference (ISI) because more symbols can contribute. This tightens the requirements of the accuracy of the clock. These narrower filters also result in more overshoot and therefore more peak carrier power. The power amplifier must then accommodate the higher peak power without distortion. The bigger amplifier causes more heat and electrical interference to be produced since the RF current in the power amplifier will interfere with other circuits. Larger, heavier batteries will be required. The alternative is to have shorter talk time and smaller batteries. Constant envelope modulation, as used in GMSK, can use class-C amplifiers which are the most efficient. In summary, spectral efficiency is highly desirable, but there are penalties in cost, size, weight, complexity, talk time, and reliability.

4.12.14.2 The Bandwidth Efficiency-Power Efficiency Trade-Off

The graph depicted in figure 4.51 shows that bandwidth efficiency is traded off against power efficiency. From the figure we can conclude that

1. The M-ary Frequency Shift Keying is power efficient, but not bandwidth efficient.
2. The M-ary Phase Shift Keying and Quadrature Amplitude Modulation are bandwidth efficient but not power efficient.

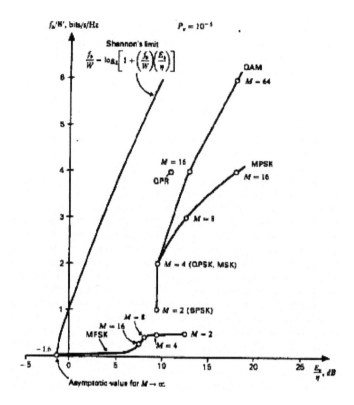

Figure 4.51 The Tradeoff Between Bandwidth Efficiency and Power Efficiency

3. Mobile radio systems are bandwidth limited, therefore Phase Shift Keying is more suited.

4.13   EYE DIAGRAMS

The eye diagram is a useful tool for the qualitative analysis of signal used in digital transmission. It provides at-a-glance evaluation of system performance and can offer insight into the nature of channel imperfections. Careful analysis of this visual display can give the user a first-order approximation of signal-to-noise, clock timing jitter, and skew.

The eye diagram is an oscilloscope display of a digital signal, repetitively sampled to get a good representation of its behavior. In a radio system, the point of measurement may be prior to the modulator in a transmitter, or following the demodulator in a receiver, depending on which portion of the system requires examination. The eye diagram can also be used to examine signal integrity in a purely digital system such as fiber optic transmission, network cables, or on a circuit board.

Impairments to the signal can occur in many places, from the pre-filtering in the transmitter, through the frequency conversion and amplifier chain, propagation path,

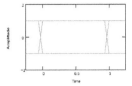 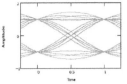

Figure 4.52 The Eye Diagrams for BPSK

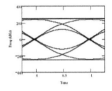 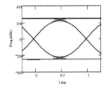 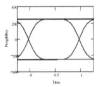

Figure 4.53 The Eye Diagrams for GMSK with BT=0.3 (left), BT=0.5 (centre) and BT=1.0 (right).

receiver front-end, IF circuits, and baseband signal processing. Information from the eye diagram can help greatly with troubleshooting. Noise problems will most often be external to the equipment and timing issues can be isolated to the receiver or transmitter with tests on each. It is also important to record the eye diagram so it is available for comparison if new problems arise in the future.

The eye diagram is also a common indicator of performance in digital transmission systems. Makers of digital communications hardware often include eye diagrams in their literature to demonstrate the signal integrity performance of their products. The eye diagrams show the signal superimposed on itself many times. If the "eye" is not "open" at the sample point, errors will occur. Eye diagram will be corrupted by noise and interference. The theoretical and practical eye diagrams of BPSK and GMSK systems are illustrated in the following Figures 4.52 and 4.53.

A final note that applies primarily to high-speed digital signal analysis is the quality of the oscilloscope used to observe the eye diagram. The bandwidth of the instrument must be sufficient to accurately display the waveform. This usually means that the bandwidth must include the third, and preferably the fifth, harmonic of the bit rate frequency.

In summary, the eye diagram is a simple and useful tool for evaluating digital transmission circuitry and systems. It provides instant visual data to verify quality or demonstrate problems. Used in conjunction with other signal integrity measurements, it can help predict performance and identify the source of system impairments.

The Communications Toolbox of MATLAB contains a function, *eyediagram(.)* to generate and plot the eye diagram.

4.13.1 PROBABILITY OF BIT ERROR

The probability of bit error (P_b) and probability of symbol error (P_s) is related to the power efficiency and the bandwidth efficiency (spectral efficiency). They are two important performance metrics in qualitatively stating the reliability of a digital communication system. To begin with, it is known that the bit error rate (BER) of a system indicates the quality of the link and:

- a BER of 10^{-3} is the maximum acceptable for voice.
- a BER of 10^{-9} is the maximum acceptable for a data link.

The ratio of the signal strength to the noise level is called the signal-to-noise ratio (SNR). If SNR is high (ie. the signal power is much greater than the noise) few errors will occur and hence BER will be low. As the SNR reduces, the noise may cause errors.

4.13.1.1 Simulation of Probability of Error Rate

In this subsection, we will take up the simulation of Probability of Error Rate on communication links, implemented using different modulation techniques. The BER results in the following sections have been produced by performing computer based simulation of the modulator, channel (AWGN), as well as bandpass filtering and demodulation. The assumption of perfect carrier and bit synchronization has been made. This is a very significant assumption and is unacceptable for a real system design, particularly at low to moderate BER. Ideally the addition of carrier and bit synchronization algorithms should be implemented in the simulator. Obviously the performance of these will vary with physical realization. A uniformly distributed random variable can be converted to a Gaussian-distributed random variable by identifying a suitable function (or mapping) that acts on the uniformly distributed random variable.

In a real mobile communications system the link between a moving (particularly vehicular) node and a base station will be subject to Rayleigh fading, resulting in fast phase shifts. This will have a significant effect on the resultant BER performance, dependent on demodulator implementation and channel, possibly increasing the required carrier-to-noise ratio (CNR) for a specific BER by as much as $10 - 15dB$.

Considerable caution must be taken when comparing graphs of BER versus E_b/N_o. Additionally, the simulations have been performed with a relatively wideband input to the demodulator. Due to the varying spectral occupancy of different modulation schemes it may be possible to employ narrower channel filtering for some schemes rather than others. For example GMSK with a BT=0.3 occupies less spectrum than MSK.

4.14 CONCLUDING REMARKS

In this chapter, we discussed the various subsystems of a typical analog or digital communication system. The signal or information passes from the source to the destination through what is called *channel*, which represents the medium that carries the signal around. Communication systems can be either *wired* or *wireless*, depending on whether there is a physical connection between the subsystems involved or not. One can also classify communication systems as *baseband* or not. An example for baseband communication system (in which there is no modulator or demodulator) is a Public Addressing system (PA system). Almost all practical communication systems contain the modulator and the demodulator, due to some specific advantages. We will consider the basic building blocks of communication systems in the following sections.

We began our discussion with an introduction to Analog Modulation schemes like Amplitude Modulation, its variants like AM-DSBSC, SSB, VSB; as well as angle modulation schemes like Frequency Modulation, and Phase Modulation. We also considered the noise performance of all the above schemes.

In subsequent sections, we discussed various digital modulation schemes such as pulse keying systems; including ASK, FSK, PSK, and their several variants; various pulse modulation systems, including Pulse Code Modulation (PCM), Differential PCM (DPCM), Adaptive Differential PCM (ADPCM), Delta Modulation (DM), Adaptive Delta Modulation (ADM), and Sigma-Delta Modulation; Quadrature Amplitude Modulation (QAM) and its variants; Minimum Shift Keying (MSK) and Gaussian MSK.

Answer to the question, which is better among PCM, DPCM, or DM, depends on the criterion used for comparison and the type of message. If one needs a relatively simple, low-cost system, DM may be the best. On the other hand, to have a high output SNR, PCM is probably the best. Now, to interface existing equipment, or otherwise to have compatibility, PCM has the advantage.

The choice of digital modulation scheme will significantly affect the characteristics, performance, and resulting physical realization of a communication system. There is no universal "best" choice of scheme, but depending on the physical characteristics of the channel, required levels of performance and target hardware trade-offs, some will prove a better fit than others. Consideration must be given to the required data rate, acceptable level of latency, available bandwidth, anticipated link budget and target hardware cost, size, and current consumption. The physical characteristics of the channel, be it hard-wired without the associated problems of fading, or a mobile, wireless communications system with fast-changing multipath, will typically significantly affect the choice of optimum system.

Communication system design requires the simultaneous conservation of bandwidth, power, and cost. In the past, it was possible to make a radio low cost by sacrificing parameters such as power and bandwidth efficiency.

Note that in Amplitude Shift Keying (ASK), pulse shaping can be employed to remove spectral spreading. ASK demonstrates poor performance, as it is heavily affected by noise and interference. Among pulse modulation schemes, Pulse Amplitude Modulation is the simplest, compared to Pulse Width Modulation and Pulse Position Modulation. But it PAM is rarely used in practice. Both PWM and PPM provide greater *noise immunity* than PAM. In PAM, additive noise directly affects the reconstructed sample value. This disruption caused by noise is less severe in PWM and PPM, where the additive noise must affect the zero crossings in order to cause an error. However, with multiplexing of channels, PWM and PPM necessitate greater care in spacing the different channels in time domain. This is a disadvantage.

Bandwidth occupancy of Frequency Shift Keying is dependent on the spacing of the two symbols. A frequency spacing of 0.5 times the symbol period is typically used. The concept of FSK can be extended to an M-ary scheme, (M-ary FSK); employing multiple frequencies as different states for different combinations of multiple bits.

Binary Phase Shift Keying (BPSK) demonstrates better performance than ASK and FSK. PSK can be expanded to an M-ary scheme, employing multiple phases and amplitudes as different states. Filtering can be employed to avoid spectral spreading.

The Phase Shift Keying (PSK) is often used, as it provides a highly bandwidth efficient modulation scheme. The Quadrature Phase Shift Keying is very robust, but requires some form of linear amplification. Alternatives like Offset QPSK and $\pi/4$-QPSK can be implemented, and reduce the envelope variations of the signal. High level M-ary schemes (such as 64-QAM, 256-QAM, 1024-QAM, 2048-QAM, or even 4096-QAM) are very bandwidth-efficient, but more susceptible to noise and require linear amplification. Constant envelope schemes (such as GMSK) can be employed since an efficient, non-linear amplifier can be used. Coherent reception provides better performance than non-coherent (or differential), but requires a more complex receiver.

The Nyquist bandwidth is the minimum bandwidth that can be used to represent a signal. It is important to limit the spectral occupancy of a signal, to improve bandwidth efficiency and remove adjacent channel interference. The raised cosine filters allow an approximation to this minimum bandwidth.

Quadrature Phase Shift Keying scheme is effectively two independent BPSK systems for Inphase and Quadrature channels (I and Q), and therefore exhibits the same performance but twice the bandwidth efficiency. Quadrature Phase Shift Keying can be filtered using raised cosine filters to achieve excellent out-of-band suppression. Large envelope variations occur during phase transitions, thus requiring linear amplification.

16-QAM has the largest distance between points, but requires very linear amplification. 16PSK has less stringent linearity requirements, but has less spacing between constellation points, and is therefore more affected by noise. M-ary schemes are more bandwidth efficient, but more susceptible to noise.

Conventional QPSK has transitions through zero ie. 180° phase transition. Highly linear amplifier is required. In Offset QPSK, the transitions on the Inphase and Quadrature channels are staggered. Phase transitions are therefore limited to 90°. In $\pi/4$-QPSK the set of constellation points are toggled each symbol, so transitions through zero cannot occur. This scheme produces the lowest envelope variations. All QPSK schemes require linear power amplifiers.

GMSK is a form of continuous-phase FSK, in which the phase is changed between symbols to provide a constant envelope. The RF bandwidth is controlled by the Gaussian lowpass filter bandwidth. The degree of filtering is expressed by multiplying the filter 3dB bandwidth by the bit period of the transmission, which is denoted as *BT*. As bandwidth of this filter is lowered the amount of intersymbol-interference introduced increases. GMSK allows efficient class C non-linear amplifiers to be used, however even with a low BT value its bandwidth efficiency is less than filtered QPSK. GMSK generally achieves a bandwidth efficiency less than 0.7 bits per second per Hz (QPSK can be as high as 1.6 bits per second per Hz). GMSK has a main lobe 1.5 times that of QPSK.

In MSK phase ramps up through 90 degrees for a binary one, and down 90 degrees for a binary zero. For GMSK transmission, a Gaussian pre-modulation baseband filter is used to suppress the high-frequency components in the data. The degree of out-of-band suppression is controlled by the BT product.

FURTHER READING

1. K.C. Raveendranathan, *Communication Systems Modelling and Simulation Using MATLAB and Simulink*, 1st Edition, Universities Press, Hyderabad, 2011.
2. B.P. Lathi, *Signal Processing & Linear Systems*, Oxford University Press, 2008.
3. Roden, Martin S., *Analog and Digital Communication Systems,* 5th Edition, Discovery Press, 2002.
4. Roddy, Dennis and John Coolen, *Electronic Communications*, 4th Edition, Pearson Education (India), 2008.
5. P.Cummiskey, N.S. Jayant, and J.L. Flanagan, Adaptive Quantization in Differential PCM Coding of Speech, *The Bell System Technical Journal,* Vol.2, No.7, September 1973.
6. William H. Tranter, K. Sam Shanmugan, Theodore S. Rappaport, and Kurt L. Kosbar, *Principles of Communication Systems Simulation with Wireless Applications*, Pearson Education, 2004.
7. Rafael C. Gonzalez, Richard E. Woods, and Steven L. Eddins, Digital Image Processing Using MATLAB, 3rd Edition, Gatesmark Publishing, 2020.

EXERCISES

1. A 50Hz, 1V peak-to-peak sinusoid modulating signal is used to pulse width modulate a 5kHz pulse train. Simulate the same using MATLAB/GNU Octave for a period of 2sec. Plot the modulated waveform and its power spectrum.
   Hints:

   a. PWM can be generated by comparing the modulating sinusoid with a biased saw tooth waveform.
   b. Alternatively, we can use the `modulate(.)` and `demod(.)` functions in MATLAB to simulated PWM modulation.
2. Simulate PAM and PPM in MATLAB/GNU Octave and plot the output waveforms choosing appropriate variables.

3. Write MATLAB/GNU Octave functions to simulate modulation and demodulation in Binary Phase Shift Keying (BPSK) of a random binary bit stream (b). The carrier frequency (fc) and symbol duration (T) should be other two input variables. The sampling frequency (fs) should be chosen appropriately.
The suggested syntax can be: $[x, nbits] = bpsk\dot{}mod(b, fc, fs, T)$ and $b = bpsk\dot{}demod(x, fc, fs, T, nbits)$; where *nbits* is the number of bits in the input binary bit stream.
Hints:
   a. For the modulator, convert the $\{0, 1\}$ bit stream to a $\{-1, +1\}$ stream and proceed further taking one bit at a time.
   b. Search MATLAB Central Website for already available code.
4. Write MATLAB/GNU Octave functions to simulate modulation and demodulation in Binary Frequency Shift Keying (BFSK) of a random binary bit stream (b). The carrier frequency (fc) and symbol duration (T) should be other two input variables. The sampling frequency (fs) should be chosen appropriately.
The suggested syntax can be: $[x, nbits] = bfsk\dot{}mod(b, fc, fs, T)$ and $b = bfsk\dot{}demod(x, fc, fs, T, nbits)$; where *nbits* is the number of bits in the input binary bit stream.
5. Plot the signal constellation of QAM and 4-PSK using appropriate MATLAB functions. What is the inference?
6. Assume that an Additive White Gaussian Noise (AWGN) having variance of 0.1 is added to a typical QAM-modulated signal. Demodulate the above signal and compute the Bit Error Ratio (BER). Develop suitable MATLAB/GNU Octave script for the above.
7. Calculate the BER for different values of Energy-per-bit to Noise power ratio $\frac{E_b}{N_o}$ for various modulation schemes and plot the performance curves.
8. Develop the MATLAB/GNU Octave script to simulate the modulation/demodulation of Pulse Code Modulation (PCM) for an arbitrary analog signal, $s(t)$.
9. Develop the MATLAB/GNU Octave script to simulate the modulation/demodulation of Delta Modulation (DM) for an arbitrary analog signal, $s(t)$.
10. Simulate Adaptive Delta Pulse Code Modulation (ADPCM) using the signal processing toolbox in MATLAB.

5 Design of Analog Filters Using MATLAB

This chapter is devoted to the design of analog filters and their simulation using MATLAB. Note that filtering of noise is an important process at various stages of a communication system, be it analog or digital. Therefore, design and simulation of filters form a major activity in maintaining signal-to-noise ratio within desired limits, in analog communication systems, so that signal reception is hassle-free.

5.1 INTRODUCTION

Filters of various types form an important subsystem in communications systems, both analog and digital. This chapter is entirely devoted to the design and simulation of analog filters of various types using appropriate built-in functions in the MATLAB Communications and Signal Processing Toolbox. We begin with the issues introduced by channel noise, and the need for filtering. Subsequently, design and simulation of Low Pass Filters, High Pass Filters, Band Pass Filters, and Band Elimination (Band Reject) are discussed. Finally, we discuss the major points to be considered while designing/simulating analog filters using appropriate toolboxes in MATLAB/GNU Octave/Mathematica or Python coding.

5.2 CHANNEL NOISE AND NEED FOR FILTERING

The communication channel or the medium used to carry the information often corrupts the message or information, which is conveyed as an electrical signal. The channel noise is to be removed at the receiver for error-free detection at the receiver so that the received signal has appreciable quality. For this purpose, filters of various configuration, with varying pass bands and stop bands are used. Note that the passband is the range of frequencies where the input signal to the filter is passed through the filter without significant attenuation. In contrast, stopband is the range of frequencies where the signal amplitude is attenuated significantly by a filter. Ideally, for a filter of any type (Low Pass, High Pass, Band Pass, or Band Reject), in the stop band the signal is attenuated to the maximum possible level (several dBs lower than at the pass band) and in the pass band attenuation is minimal, close to $0 dB$. Also, it is important to note that response of any filter is far from ideal in practice. There is also a significant amount of ripple (slight variations in response or attenuation level) in the pass band termed as *Pass Band Ripple*.

The channel noise has often been modeled as Additive White and Gaussian Noise (AWGN), meaning it adds on to the signal transmitted; spreads over the entire spectrum of frequencies ("White"); and has a Gaussian Probability Density Function

DOI: 10.1201/9781003527589-5

(PDF). The channel is therefore termed as an AWGN channel. We will now consider various types of filters with different attenuation characteristics/responses.

Additive White Gaussian Noise (AWGN) is a common model for communication channels because it simplifies the mathematical analysis and allows for the application of powerful tools from probability theory and statistics. There are several reasons why the AWGN model is a suitable model for communication channels:

- Simplicity and Easiness to Analyze: The AWGN model simplifies the channel modeling process and enables the use of mathematical tools like linear algebra and probability theory. This simplification allows for a more straightforward analysis of communication system performance.
- Central Limit Theorem: The Central Limit Theorem states that the sum (or average) of a large number of independent, identically distributed random variables (i.i.d. random variables) converges to a Gaussian (normal) distribution. In communication systems, many independent noise sources (thermal noise, interference, etc.) contribute to the overall channel noise, making the assumption of channel being AWGN reasonable to model, correctly.
- Realistic Noise Model: The AWGN model captures the essence of noise in many communication systems, where noise is often introduced by thermal effects in electronic components and other random processes. Gaussian noise model is often used to represent the cumulative effect of many small, independent noise sources.
- Ease of Performance Analysis: The Gaussian distribution is completely characterized by its mean and variance, simplifying the analysis of system performance metrics such as bit error rate (BER) or signal-to-noise ratio (SNR). This makes it easier to predict and optimize the behavior of communication systems.
- Mathematical Convenience: Gaussian noise is mathematically convenient to work with. Many mathematical operations, such as convolution and linear filtering, preserve the Gaussian distribution, making it easier to analyze the impact of noise on signals.

However, while the AWGN model is a simplification and may not perfectly represent every real-world communication channel, it serves as a useful starting point for system design, analysis, and optimization. Communication system designers can later incorporate more complex/more appropriate channel models or adapt their designs based on the specific characteristics of the actual channel. There are various other channel noise models used in communication systems modeling to capture different characteristics and challenges present in real-world channels. Some of them are:

1. Rayleigh Fading Channel Model: This model accounts for the effects of multipath propagation, where a signal reaches the receiver through multiple paths with different delays and amplitudes. This model is applied commonly in wireless communication systems where signals experience reflections and scatterings.

2. Rician Fading Channel Model: Though it is similar to Rayleigh fading model, but it comes with an additional line-of-sight component. It is suitable for channels where there is a dominant direct path in addition to scattered paths. This is used in environments with a clear line of sight, such as some wireless communication systems.

3. Frequency-Selective Fading Channel Model: This model takes into account frequency-dependent fading, where different frequency components of the signal experience different fading effects. It is applicable in high-mobility communication, vehicular communication systems.

4. Impulsive Noise Channel Model: It models sudden and short-duration bursts of interference or noise; and it is applicable to channels where sporadic interference, like impulse noise from electrical devices, can affect the communication channel.

5. Non-Gaussian Noise Models: In addition to AWGN, other non-Gaussian noise models may be considered, such as Laplace noise or Poisson noise models, depending on the specific characteristics of the noise source. These are used when the noise distribution deviates significantly from a Gaussian distribution.

6. Channel with Memory: This noise model is applicable to channels where the symbols received have a dependency on previous symbols, introducing temporal correlation; and it is suitable for channels where the channel exhibits memory effects, such as fading channels with memory.

7. Quantization Noise Model: In this model, the noise arises from the quantization process in analog-to-digital and digital-to-analog converters. This noise model is applicable in digital communication systems involving signal processing and conversion between analog and digital domains.

Ultimately, channel noise models are chosen based on the specific characteristics/challenges of the communication environment. As a concluding note we may add that the choice of a particular model depends on the application, the nature of the communication channel, and the desired level of accuracy in the analysis.

5.3   DESIGN AND SIMULATION OF LOW PASS FILTERS (LPF)

A Low Pass Filter has a pass band spanning over the lower frequencies (it almost passes lower frequencies with minimal attenuation) and attenuates heavily the higher frequencies beyond the so-called cut-off frequency. Often there is a passband ripple visible in the passband which is one of the important design criteria. Depending on the frequency response characteristics needed for a specific application, which includes the passband ripple, the width of the transition band, and the stopband attenuation, we can design and simulate filters with different response types. These include *Butterworth, Chebyshev, Bessel, and Elliptic filters.* Each one of them is characterized by a typical response characteristics. For example, a low pass filter of Butterworth type shows an almost flat passband response (in other words with minimal passband ripple); and a wider transition band, and stopband with moderate attenuation. In the following subsection, we will discuss a typical low-pass filter design using the MATLAB Signal Processing Toolbox.

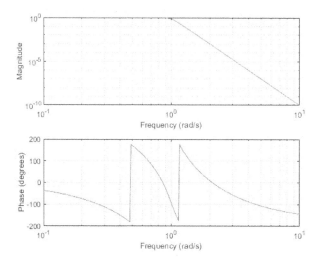

Figure 5.1   The Total Frequency Response Plot of a Order-10 Butterworth Low Pass Filter

5.3.1   ANALOG FILTER DESIGN AND SIMULATION USING SIGNAL PROCESSING TOOLBOX

The Signal Processing Toolbox provides functions to design and analyze analog filters, including Butterworth, Chebyshev, Bessel, and elliptic designs. It is also possible to perform analog-to-digital filter conversion using discretization methods such as impulse invariance and the bilinear transformation. In GNU Octave there are compatible functions for the same purpose.

5.3.1.1   Design of a Typical Butterworth LPF

The MATLAB Signal Processing Library function buttap can be used to obtain a Butterworth analog filter prototype. The exact syntax of the function is $[z, p, k] = buttap(n)$ where n is the order; z is the zeros, p is the poles and k is the gain. Along with this, we have to use zp2tf() to get the transfer function of the analog filter prototype. A sample script is appended below to design a 10th-order analog LPF prototype and plot its frequency response.

```
[z,p,k] = buttap(10);         % Butterworth filter prototype
[num,den] = zp2tf(z,p,k);     % Convert to transfer function form
freqs(num,den)                % Plot the frequency response.
```

The total frequency response plot of the above MATLAB/GNU Octave script is shown in figure 5.1. The MATLAB/GNU Octave function buttord(.)[1] computes

---

[1]In GNU Octave, the command pkg load signal should be given at command prompt prior to evoking all the signal processing library functions

the order of the Butterworth filter for a given set of parameters like passband-edge frequency, stopband-edge frequency, allowable passband ripple in dBs, and the minimum attenuation in the stopband in dBs. We can combine the buttord(.) function with buttap(.) to design analog filters of Butterworth type, given appropriate specifications as mentioned above. For example, the command $[N, Wn] = buttord(Wp, Ws, Rp, Rs)$ returns the order N of the lowest order Butterworth filter which has a passband ripple of no more than Rp dB and a stopband attenuation of at least Rs dB. Wp and Ws are the passband and stopband edge frequencies, normalized from 0 to 1; where 1 corresponds to π radians/sample. Also, for a lowpass filter, we need to ensure that $Ws > Wp$.

Similarly, for designing lowpass filters of other genre like Chebyshev Type I and II, Bessel, and Elliptical there are similar function pairs like cheby1ord, cheby1ap, cheby2ord, cheby2ap, besselord, besselap, ellipord, and ellipap. A typical MATLAB script to design, simulate, and plot various types of low-pass filters (to compare their performance) is given below:

```
%% MATLAB Script to Design a 5th-order analog Butterworth lowpass
% filter with a cutoff frequency of 4MHz. Multiply by $2\Pi$ to convert the
% frequency to radians per second. Compute the frequency response of
% the filter at 4096 points.
n = 5; % Order of the Butterworth LPF = 5
f = 4e6;%Cutoff frequency = 4MHz
% To design an analog filter parameter 's' is used
[zb,pb,kb] = butter(n,2*pi*f,'s');
[bb,ab] = zp2tf(zb,pb,kb);
[hb,wb] = freqs(bb,ab,4096);
%Design a 5th-order Chebyshev Type I filter with the same edge
% frequency  and 3 dB of passband ripple. Compute its frequency response.
[z1,p1,k1] = cheby1(n,3,2*pi*f,'s');
[b1,a1] = zp2tf(z1,p1,k1);
[h1,w1] = freqs(b1,a1,4096);
%Repeat for a 5th-order Chebyshev Type II filter with the same edge
% frequency and 30 dB of stopband attenuation
[z2,p2,k2] = cheby2(n,30,2*pi*f,'s');
[b2,a2] = zp2tf(z2,p2,k2);
[h2,w2] = freqs(b2,a2,4096);
% Repeat for 5th-order Elliptic filter with same parameters as above
[ze,pe,ke] = ellip(n,3,30,2*pi*f,'s');
[be,ae] = zp2tf(ze,pe,ke);
[he,we] = freqs(be,ae,4096);
% Plot the attenuation in decibels. Express the frequency in MHz
% and  compare the performance of all four filters.
plot(wb/(f*pi),mag2db(abs(hb)),'Color',[0,0,0],LineWidth=1.0)
hold on
plot(w1/(f*pi),mag2db(abs(h1)),'Color',[0,0,0],LineWidth=1.5)
plot(w2/(f*pi),mag2db(abs(h2)),'Color',[0,0,0],LineWidth=2.0)
plot(we/(f*pi),mag2db(abs(he)),'Color',[0,0,0],LineWidth=2.5)
axis([0 8 -40 5]); grid;hold off
xlabel('Frequency  in MHz')
```

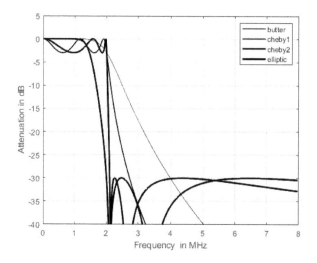

Figure 5.2   Comparison of Performances of Low Pass Filters

```
ylabel('Attenuation in dB')
legend('butter','cheby1','cheby2','elliptic')
```

The output of the simulation is shown in figure 5.2. We can make the following observations from the response plots.

1. The Butterworth and Chebyshev Type II filters have flat passbands and wide transition bands.
2. The Chebyshev Type I and elliptic filters roll off faster but have passband ripple.
3. The frequency input to the Chebyshev Type II design function sets the beginning of the stopband rather than the end of the passband.

5.4   DESIGN AND SIMULATION OF HIGH PASS FILTERS (HPF)

The design and simulation of high-pass filters using MATLAB Signal Processing Toolbox is similar to that of the low-pass filters explained in the previous section. We will develop a similar MATLAB script to compare the performance of various types of analog high pass filters of order 7 and stopband edge frequency, $4MHz$. The code is appended below and the output is shown in figure 5.3.

```
%% MATLAB Script to Design a 7th-order analog Butterworth highpass
% filter with a cutoff frequency of 4MHz. Multiply by 2Pi to convert the
% frequency to radians per second. Compute the frequency response of
% the filter at 4096 points.
n = 7; % Order of the Butterworth LPF = 7
f = 4e6;%Cutoff frequency = 4MHz
```

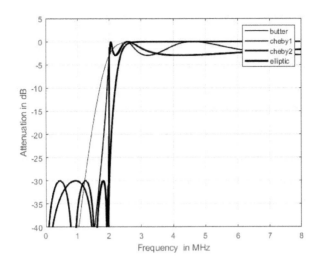

Figure 5.3   Comparison of Performances of High Pass Filters

```
% design an analog filter parameter 's' is used
[zb,pb,kb] = butter(n,2*pi*f,'high','s');
[bb,ab] = zp2tf(zb,pb,kb);
[hb,wb] = freqs(bb,ab,4096);
%Design a 7th-order Chebyshev Type I filter with the same edge
% frequency  and 3 dB of passband ripple. Compute its frequency response.
[z1,p1,k1] = cheby1(n,3,2*pi*f,'high','s');
[b1,a1] = zp2tf(z1,p1,k1);
[h1,w1] = freqs(b1,a1,4096);
%Repeat for a 7th-order Chebyshev Type II filter with the same edge
% frequency and 30 dB of stopband attenuation
[z2,p2,k2] = cheby2(n,30,2*pi*f,'high','s');
[b2,a2] = zp2tf(z2,p2,k2);
[h2,w2] = freqs(b2,a2,4096);
% Repeat for 7th-order Elliptic filter with same parameters as above
[ze,pe,ke] = ellip(n,3,30,2*pi*f,'high','s');
[be,ae] = zp2tf(ze,pe,ke);
[he,we] = freqs(be,ae,4096);
% Plot the attenuation in decibels. Express the frequency in MHz
% and  compare the performance of all four filters.
plot(wb/(f*pi),mag2db(abs(hb)),'Color',[0,0,0],LineWidth=1.0)
hold on
plot(w1/(f*pi),mag2db(abs(h1)),'Color',[0,0,0],LineWidth=1.5)
plot(w2/(f*pi),mag2db(abs(h2)),'Color',[0,0,0],LineWidth=2.0)
plot(we/(f*pi),mag2db(abs(he)),'Color',[0,0,0],LineWidth=2.5)
axis([0 8 -40 5]); grid;hold off
xlabel('Frequency  in MHz')
ylabel('Attenuation in dB')
legend('butter','cheby1','cheby2','elliptic')
```

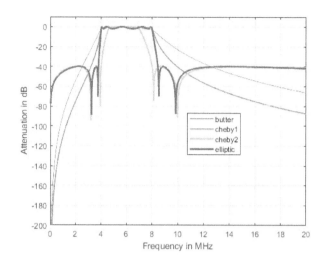

Figure 5.4 Comparison of Performances of Band Pass Filters

5.5   DESIGN AND SIMULATION OF BAND-PASS FILTERS (BPF)

In the case of band-pass filters, if one examines the pass band, it is spread in between two stopbands. Using the library functions of MATLAB Signal Processing Toolbox we can easily design and simulate band-pass filters of various types. Band pass filters are used to pass on a selected band of frequencies within the signal being transmitted where as band of frequencies on either side of the pass band is suppressed considerably. Thus, the bandpass filter has a passband embedded between two stopbands.

A typical MATLAB script to design and simulate bandpass filters of various types and to compare their performances is given below. The resulting output of the simulation is shown in figure 5.4.

```
%% MATLAB Script to Design a 5th-order analog Butterworth Bandpass
% filter with a lower cutoff frequency of 4MHz, Upper Cutoff frequency
% 8MHz and pass band in between. The minimum attenuation in the
% stop band is 40dB and the maximum ripple in the passband is 2dB.
% Multiply by 2Pi to convert the frequency to radians per second.
% Compute the frequency response of the filter at 4096 points.
clear all; close all; clf;
n = 5; % Order of the Butterworth BPF = 5
fl = 4e6; % Lower Cutoff frequency = 4MHz
fu = 8e6; % Upper Cutoff frequency = 8MHz
fs=16e6; % Sampling frequency; fs>= 2*max(fl,fu).
fp = 2 * pi * [fl, fu] / fs; % Normalized Passband;
Rp = 2; % Maximum allowable passband ripple = 2dB.
Rs = 40; % Minimum Stopband Attenuation = 40dB.

% Design an analog Butterworth filter
```

```
[zb, pb, kb] = butter(n, fp, 'bandpass', 's');
[bb, ab] = zp2tf(zb, pb, kb);
[hb, wb] = freqs(bb, ab, 4096);

% Design a 5th-order Chebyshev Type I filter
[z1, p1, k1] = cheby1(n, Rp, fp, 's');
[b1, a1] = zp2tf(z1, p1, k1);
[h1, w1] = freqs(b1, a1, 4096);

% Repeat for a 5th-order Chebyshev Type II filter
[z2, p2, k2] = cheby2(n, Rs, fp, 's');
[b2, a2] = zp2tf(z2, p2, k2);
[h2, w2] = freqs(b2, a2, 4096);

% Repeat for 5th-order Elliptic filter
[ze, pe, ke] = ellip(n, Rp, Rs, fp, 's');
[be, ae] = zp2tf(ze, pe, ke);
[he, we] = freqs(be, ae, 4096);

% Plot the attenuation in decibels. Express the frequency in MHz
% and  compare the performance of all four filters.
plot(wb / (2 * pi) * fs / 1e6, mag2db(abs(hb)), 'LineWidth', 1.0);
hold on;
plot(w1 / (2 * pi) * fs / 1e6, mag2db(abs(h1)), 'LineWidth', 1.5)
plot(w2 / (2 * pi) * fs / 1e6, mag2db(abs(h2)), 'LineWidth', 2.0)
plot(we / (2 * pi) * fs / 1e6, mag2db(abs(he)), 'LineWidth', 2.5)
grid; hold off; axis([0 20 -200 10])
xlabel('Frequency in MHz'); ylabel('Attenuation in dB');
legend('butter', 'cheby1', 'cheby2', 'elliptic');
```

The following points are to be noted while designing and simulating analog band-pass and band stop filters, regarding the pass band edge frequencies, denoted by Wp radians/seconds:

1. The passband edge frequency, is specified as a two-element vector. The passband edge frequency is the frequency at which the magnitude response of the filter is Rp decibels. Smaller values of passband ripple, Rp, result in wider transition bands.
2. If Wp is a scalar, then *cheby1(.)* designs a lowpass or highpass filter with edge frequency Wp.
3. If Wp is the two-element vector $[w1\ w2]$, where $w1 < w2$, then *cheby1* designs a bandpass or bandstop filter with lower edge frequency $w1$ and higher edge frequency $w2$.
4. For analog filters, the passband edge frequencies must be expressed in radians per second and can take on any positive value.

5.5.1   FILTER VISUALIZATION TOOL, FVTOOL(.)

The *fvtool(.)* library function in Signal Processing Toolbox of MATLAB can help us to visualize easily the analog or digital filter we have designed. For example, to

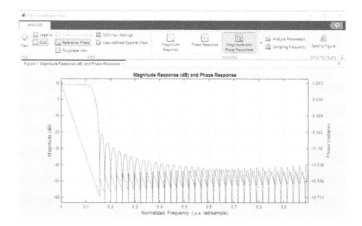

Figure 5.5   The Filter Visualization Tool Window

create a *filter object* for a Square Root Raised Cosine (SRRC) Transmit Filter with
unity passband gain (i.e. 0*dB*); the following code is sufficient:

```
txfilter = comm.RaisedCosineTransmitFilter; % evoke the appropriate function from..
%% Communication Systems Toolbox
fvtool(txfilter);
```

The resulting output window is given in figure 5.5 As another example, we can
use the following MATLAB script to design a 5th order Butterworth filter with lower
cutoff frequency of 4GHz and upper cutoff frequency of 8GHz.

```
% Design an analog 5th Butterworth filter
% and use fvtool to visualize the response
n=5; % order of the filter.
fl=4; % lower edge frequency=4GHz
fu=8; % upper edge frequency=8GHz
fs=16; % Sampling frequency=16GHz
fp=2*pi*[fl,fu]; % denormalize the passband edge frequencies
[zb, pb, kb] = butter(n, fp, 'bandpass', 's');
[bb, ab] = zp2tf(zb, pb, kb);
% Visualize the filter response using fvtool
fvtool(bb, ab, 'FrequencyRange', 'Specify freq. vector',...
    'FrequencyVector', linspace(0, fs/2, 4096));
```

The output of the script is illustrated in figure 5.6

5.5.2   FILTER DESIGNER APP

The Filter Design and Analysis Tool (fdatool) was yet another tool in MATLAB and
it was replaced by Filter Designer App in later versions. The Filter Designer App

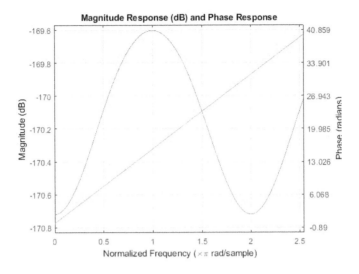

Figure 5.6 Design Visualization of a 5th Order Butterworth Filter Using fvtool.

available in DSP System Toolbox of MATLAB enables us to design and analyze digital filters. We can also import and modify existing filter designs. Using the app, we can:

1. Choose a response type and filter design method
2. Set filter design specifications
3. Analyze, edit, and optimize a filter design
4. Export a filter design or generate MATLAB code

There are two ways to evoke the Filter Designer app.

- At the MATLAB Toolstrip, on the Apps tab, under Signal Processing and Communications, click the app icon. Or,
- Enter **filterDesigner** in the MATLAB command prompt.

After opening the filter designer app, we can choose the parameters of the digital filter as desired. These include the Response type (Low Pass, High Pass, Band Pass, Band Stop, and Differentiator-all these have several variants), Design Method (IIR or FIR with different variants), Filter Order, Frequency Specifications, and Magnitude Specifications. See MATLAB online documentation for more details.

As an example, choose the following options for a digital FIR bandpass, equiripple, elliptic filter:

- Filter order = 10;
- Sampling Frequency, Fs = 48kHz, Lower Stop Band Frequency, Fstop1 = 7.2kHz, Passband Edge Frequency, Fpass1 = 9.6kHz, Fpass2 = 12kHz, Upper Stop Band Frequency, Fstop2 = 14.4kHz.

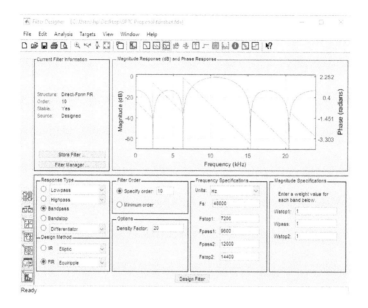

Figure 5.7   Visualization of a 10th-order FIR Band Pass Elliptic Filter

- All specifications under Magnitude Specifications, viz. Wstop1, Wpass, and Wstop2 are made 1.

Then, click on Design Filter button shown on the bottom of the window and further, click on the **Magnitude and Phase Responses** icon at the top of the window. The app will generate the magnitude and phase response of the bandpass filter. It is shown in figure 5.7.

Note that the FIR filter has a linear phase response.

5.6   DESIGN AND SIMULATION OF BAND ELIMINATION (BAND REJECT) FILTERS (BRF)

The design and simulation of Band Reject Filter (or Band Elimination/Band Stop filter) is similar to that of Band Pass Filter. The MATLAB script to design and plot the responses of band-reject filters are appended:

```
%% MATLAB Script to Design a 5th-order analog Butterworth Bandstop
% filter with a lower cutoff frequency of 4MHz, Upper Cutoff frequency
% 10MHz. The minimum attenuation in the stop band is 40dB and
% the maximum ripple in the passband is 2dB.
% Multiply by 2Pi to convert the frequency to radians per second.
% Compute the frequency response of the filter at 4096 points.
clear all; close all; clf;
n = 5; % Order of the Butterworth BSF = 5
fl = 4e6; % Lower Cutoff frequency = 4MHz
```

```
fu = 10e6; % Upper Cutoff frequency = 10MHz
fs=20e6; % Sampling frequency; fs>= 2*max(fl,fu).
fp = 2 * pi * [fl, fu] / fs; % Normalized Passband;
Rp = 2; % Maximum allowable passband ripple = 2dB.
Rs = 40; % Minimum Stopband Attenuation = 40dB.


% Design an analog Butterworth band stop filter
[zb, pb, kb] = butter(n,fp, 'stop', 's');
[bb, ab] = zp2tf(zb, pb, kb);
[hb, wb] = freqs(bb, ab, 4096);


% Design a 5th-order Chebyshev Type I filter
[z1, p1, k1] = cheby1(n, Rp, fp, 'stop', 's');
[b1, a1] = zp2tf(z1, p1, k1);
[h1, w1] = freqs(b1, a1, 4096);


% Repeat for a 5th-order Chebyshev Type II filter
[z2, p2, k2] = cheby2(n, Rs, fp,'stop', 's');
[b2, a2] = zp2tf(z2, p2, k2);
[h2, w2] = freqs(b2, a2, 4096);


% Repeat for 5th-order Elliptic filter
[ze, pe, ke] = ellip(n, Rp, Rs, fp,'stop', 's');
[be, ae] = zp2tf(ze, pe, ke);
[he, we] = freqs(be, ae, 4096);
% Plot the attenuation in decibels. Express the frequency in MHz
% and  compare the performance of all four filters.
plot(wb / (2 * pi) * fs / 1e6, mag2db(abs(hb)), 'LineWidth', 0.5);
hold on;
plot(w1 / (2 * pi) * fs / 1e6, mag2db(abs(h1)), 'LineWidth', 1)
plot(w2 / (2 * pi) * fs / 1e6, mag2db(abs(h2)), 'LineWidth', 2)
plot(we / (2 * pi) * fs / 1e6, mag2db(abs(he)), 'LineWidth', 3)
grid; hold off;axis([0 20 -200 10])
xlabel('Frequency in MHz'); ylabel('Attenuation in dB');
legend('butter', 'cheby1', 'cheby2', 'elliptic');
```

The result of the simulation is illustrated in figure 5.8.

5.7   DESIGN AND SIMULATION OF COMB FILTERS

A comb filter is a type of filter that selectively amplifies or attenuates periodic components in a signal. It is called a "comb" filter because its frequency response resembles the teeth of a comb. Comb filters are commonly used in various signal processing applications, including audio processing, echo cancellation, and modulation/demodulation schemes.

The frequency response of a comb filter has peaks and notches at regularly spaced intervals. The general transfer function of a digital comb filter is given by:

$$H(z) = 1 + \alpha z^{-M} \tag{5.7.1}$$

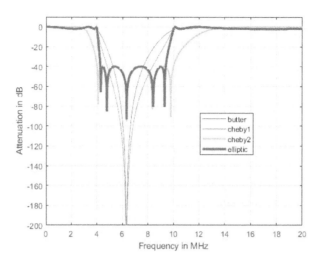

Figure 5.8 Comparison of Performances of Band Stop Filter

where α is the comb filter coefficient and M is the delay length. The choice of α and M determines the characteristics of the comb filter. A typical MATLAB code to design and visualize a digital comb filter is given below:

```
%% MATLAB Script to design & Visualize Comb Filter
% Comb filter design parameters
alpha = 0.9;   % Comb filter coefficient
M = 10;        % Delay length
% Comb filter transfer function coefficients
b = [1, zeros(1, M), alpha]; % Numerator Coefficients
a = 1; % denominator Coefficient
% Frequency response visualization
fvtool(b, a, 'Analysis','freq');
title('Comb Filter Frequency Response');
```

The response of the Comb filter is shown in figure 5.9

5.7.1 ANALOG COMB FILTER

The transfer function of an nth-order analog comb filter can be expressed in the Laplace domain as follows:

$$H(s) = \frac{1}{(1 - Ge^{-Ts})^n} \qquad (5.7.2)$$

where $H(s)$ is the transfer function; G is the feedback-gain; and T is the Time Delay. The following MATLAB code snippet can design and simulate a 40th-order analog comb filter. Figure 5.10 shows its response.

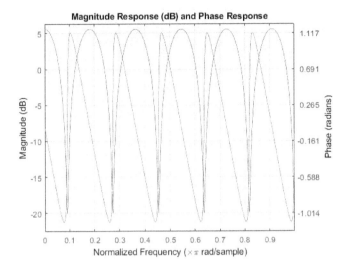

Figure 5.9   Response of a Comb Filter

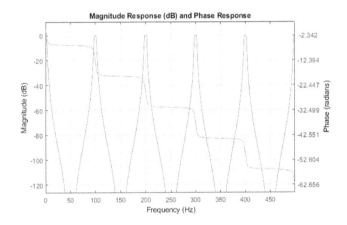

Figure 5.10   Response of an Analog 40th-Order Comb Filter

```
%% MATLAB code snippet to design and plot a analog Comb filter.
%% parameters below:
% peak - specifies the peak response of the comb filter.
% L -   determines the number of peaks or notches, which are
% equally spaced over the normalized frequency interval [-1,1].
% BW - Bandwidth of the notch or peak. By default the bandwidth
% is calculated at the point {3 dB down from the center frequency.
% GBW - Gain at which the bandwidth is measured.
% Nsh - Shelving filter order. Nsh represents a positive integer
% that determines the sharpness of the peaks or notches.
```

```
% The greater the value of the shelving filter order, the steeper the
% slope of the peak or notch. This results in a filter of order L*Nsh.
d = fdesign.comb('peak','L,BW,GBW,Nsh',10,5,-4,4,1e3);
% design a peak comb filter with L=10, BW=5
% GBW=-4dB, Nsh=4; and frequency =1kHz
Hd=design(d,'SystemObject',true);
% Obtains a comb filter handle object.
fvtool(Hd); % Obtains the response.
```

5.8 CONCLUDING REMARKS

In this chapter, we studied various noise models that corrupted the signals while they traverse the channel. We began our discussion by introducing the most common noise model used to model the noise that was added to the signal at the channel such as AWGN. We considered the reasons for the AWGN model being the most popular noise model. We also considered other channel-noise models like Rayleigh Fading Channel Model, Rician Fading Channel Model, Frequency-Selective Fading Channel Model, Impulsive Noise Fading Channel Model, Major Non-Gaussian Channel Models (like Laplace Noise and Poisson Noise Models), Noise Models for channels with Memory, and finally Quantization Noise Model. Unlike others, the quantization noise results from the Analog-to-digital conversion process encountered in the digital signal processing systems. Subsequently, in this chapter, we also discussed the design and simulation of low pass, high pass, band pass, band stop, and comb filters using MATLAB. All the relevant MATLAB library functions in Signal Processing, DSP System, and Communication Toolboxes were also introduced.

FURTHER READING

1. Luis F. Chaparro, *Signals and Systems Using MATLAB*, Cengage Learning.
2. Robert J. Schilling and Sandra L. Harris, *Introduction to Digital Signal Processing Using MATLAB*, 2nd Edition.
3. B.P. Lathi, *Signal Processing & Linear Systems*, Oxford University Press, 2008.
4. Dennis Roddy and John Coolen, *Electronic Communications*, 4th Edition, Pearson Education (India), 2008.
5. Ali Grami, *Introduction to Digital Communications*, Elsevier, 2016.
6. William H. Tranter, K. Sam Shanmugan, Theodore S. Rappaport, and Kurt L. Kosbar, *Principles of Communication Systems Simulation with Wireless Applications*, Pearson Education, 2004.
7. Rafael C. Gonzalez, Richard E. Woods and Steven L. Eddins, Digital Image Processing Using MATLAB, 3rd Edition, Gatesmark Publishing, 2020.
8. The MathWorks Inc., *DSP System Toolbox Reference*, 2023.
9. The MathWorks Inc., *DSP System Toolbox User Guide*, 2023.
10. The MathWorks Inc., *Communications Toolbox Reference*, 2023.
11. The MathWorks Inc., *Signal Processing Toolbox User Guide*, 2023

EXERCISES

1. Design, simulate, and plot the magnitude and phase response of a Butterworth analog low pass filter with the following specifications: cut-off frequency=60Hz, maximum pass band ripple = 0.1dB, and minimum stop band attenuation=40dB. Decide the order of the filter using the MATLAB function, *buttord(.)* and proceed. What is your observation?

2. Repeat exercise 1 with the following modifications: design a Chebychev Type I analog high pass filter, with passband edge frequency 2*kHz*, pass band ripple=2dB, and stop band attenuation=60dB.

3. Write MATLAB/GNU Octave script to design, simulate, and plot the responses of a Butterworth, Chebyshev Type I & II, and Elliptic analog low pass filter with the following specifications: cut-off frequency=60kHz; maximum pass band ripple = 1dB; minimum stop band attenuation = 60dB.

4. Design, simulate, and plot the amplitude and frequency response of an analog high pass filter with following specifications: pass band edge frequency=2MHz; maximum pass band ripple=0.2dB; minimum stop band attenuation=60dB. Filter type: elliptic.

5. Design, simulate, and plot the response of an analog band pass filter with following specification using MATLAB/GNU Octave: pass band ripple = 1dB; pass band edge frequencies: 6MHz and 10MHz; minimum attenuation in stop band = 60dB; Filter type: Butterworth, Chebyshev Type II. What is the inference?

6. Assume that an Additive White Gaussian Noise (AWGN) having variance of 0.1 is added to a typical sinusoidal signal of amplitude 2V-pp and frequency 1kHz. De-noise the above signal using a suitable analog low-pass filter Develop suitable MATLAB/GNU Octave script for the above.

7. Design, simulate, and plot the response of an analog band stop filter with following specification using MATLAB/GNU Octave: pass band ripple = 2dB; pass band edge frequencies: 8GHz and 12GHz; minimum attenuation in stop band = 60dB; Filter type: Chebyshev Type I and II; and Elliptic. What is the inference?

8. Design, simulate, and plot the response of an analog band-pass filter with following specification using MATLAB/GNU Octave: pass band ripple = 1dB; pass band edge frequencies:10MHz and 16MHz; minimum attenuation in stop band = 60dB; Filter type: Chebyshev Type I and II; and Elliptic.

9. Design, simulate, and plot a comb filter with following parameters: order N=30; response type = 'notch'; Number of peaks =5; Band Width = 20; Gain at bandwidth=-6dB, Shelving filter order=6, and frequency=600Hz.

10. Design, simulate, and plot a comb filter with following parameters: order N=20; response type = 'peak'; Number of peaks =5; Band Width = 10; Gain at bandwidth=-6dB, Shelving filter order=4, and frequency=800Hz.

6 Design of Digital Filters

In this chapter, we discuss the design and simulation of digital filters using the Signal Processing, Communications, and DSP System Toolboxes of MATLAB. Note that filtering of noise is an important process at various stages of a communication system, be it analog or digital. Therefore, design and simulation of filters form a major activity in maintaining signal-to-noise ratio within desired limits, in digital communication systems, so that signal reception is hassle-free and of desired quality. Note that an improvement in SNR results in a better (i.e. lower in value) of Bit Error Rate (BER) at the receiver side of digital communication system.

6.1 INTRODUCTION

Filters of various types form an important subsystem in communications systems. This chapter is entirely devoted to the design and simulation of digital filters of various types using appropriate built-in functions in the Communications, DSP System, and Signal Processing Toolboxes in MATLAB. When we process the signals in the digital domain, there is a need to convert the analog signal to digital format using appropriate analog-to-digital converters (ADCs). The analog-to-digital conversion is an approximation process and results in the *quantization noise* as mentioned in chapter 4 of this book. We begin with a brief discussion of the issues introduced by quantization noise, and the need for filtering. Subsequently, the design and simulation of Low Pass Filters, High Pass Filters, Band Pass Filters, and Band Elimination (Band Reject), All Pass Filter, Comb Filters, Moving Average Filters, Wavelet denoising, and Wiener Filtering are discussed. Finally, we discuss the major points to be considered while designing/simulating digital filters using appropriate toolboxes in MATLAB/GNU Octave/Mathematica or Python.

6.2 MITIGATION OF QUANTIZATION NOISE

Quantization noise is a phenomenon that arises in digital signal processing and data representation when analog signals are converted into digital form through a process known as quantization. This conversion involves representing continuous analog signals with discrete digital values, leading to a loss of precision. The discrepancy between the original analog signal and its digitized representation introduces quantization error, commonly referred to as quantization noise.

Quantization is an essential step in various digital systems, including audio and image processing, telecommunications, and data compression. The precision of the quantization process is determined by the number of bits used to represent each sample; with higher the number of bits providing greater precision.

The quantization process introduces errors because it forces each sample to be represented by the nearest available discrete level. The difference between the actual

DOI: 10.1201/9781003527589-6

analog value and its quantized representation is the quantization error. This error manifests as an unwanted and often audible or visible noise in the processed signal.

In the context of quantization, the signal represents the original analog signal, and the noise is the quantization error. As the number of bits used in the quantization increases, providing more levels for representation, the quantization noise decreases, resulting in a higher SNR and better signal quality. This metric is often denoted as Signal to Quantization Noise Ratio (SQNR). For a uniform quantizer, where the quantization levels are equally spaced, the root mean square (RMS) value of the quantization error is inversely proportional to the square root of the number of quantization levels. This relationship is described by the formula:

$$RMS_{error} = \frac{Q}{\sqrt{12}}$$

where Q is the quantization step size. Despite efforts to minimize quantization noise by increasing bit depth, practical constraints often limit the number of bits used in digital systems. This trade-off between precision and resource constraints is a fundamental consideration in the design of digital systems.

6.2.1 MITIGATION OF QUANTIZATION ERROR

Various techniques have been developed to mitigate the impact of quantization noise. Dithering is one such method that involves adding a small amount of random noise before quantization. This additional noise helps to distribute the quantization error more evenly, reducing the audibility or visibility of the noise in the processed signal.

We can reduce the quantization noise by oversampling and filtering. Oversampling is the process of sampling the signal at a higher rate than the Nyquist rate, to reduce the quantization error and the aliasing effect. Filtering is the process of applying a low-pass filter to the signal before or after sampling, to remove the unwanted frequency components and noise.

In conclusion, increasing the sampling rate and the number of bits per sample can reduce the quantization error/noise by increasing the resolution and precision of the digital signal. We can also use low-pass filters to mitigate the quantization noise.

A sample python code snippet is given below to simulate the effect of quantization noise; and its denoising processing using a Butterworth low pass filter:

```
import numpy as np
import matplotlib.pyplot as plt
from scipy import signal

# Create a sample signal
t = np.linspace(0, 1, 1000, endpoint=False)
x = np.sin(2 * np.pi * 5 * t)

# Simulate quantization noise
n_bits = 16  # Number of bits for quantization
```

```
x_quantized = np.round(x * (2**(n_bits-1))) / (2**(n_bits-1))
quantization_noise = x - x_quantized

# Plot the original signal and quantization noise
plt.figure()
plt.subplot(3, 1, 1)
plt.plot(t, x, label='Original Signal')
plt.plot(t, x_quantized, label='Quantized Signal')
plt.xlabel('Time')
plt.ylabel('Amplitude')
plt.legend()

plt.subplot(3, 1, 2)
plt.plot(t, quantization_noise, label='Quantization Noise')
plt.xlabel('Time')
plt.ylabel('Amplitude')
plt.legend()

# De-noising using a low-pass filter
cutoff_freq = 10  # Cutoff frequency for the low-pass filter
nyquist_freq = 0.5 * 1000  # Nyquist frequency
normalized_cutoff_freq = cutoff_freq / nyquist_freq
b, a = signal.butter(4, normalized_cutoff_freq, btype='low', analog=False)
x_denoised = signal.filtfilt(b, a, x_quantized)

# Plot the original signal, quantized signal, and de-noised signal
plt.subplot(3,1,3)
plt.plot(t, x, label='Original Signal')
plt.plot(t, x_quantized, label='Quantized Signal')
plt.plot(t, x_denoised, label='De-noised Signal')
plt.xlabel('Time')
plt.ylabel('Amplitude')
plt.legend()
plt.savefig('fig6_1.jpg')
plt.show()
```

The output of the simulation is given in figure 6.1

6.3   DIGITAL FIR AND IIR FILTERS

Digital filters are essential components in digital signal processing, used to modify or analyze signals in various applications such as audio processing, communication systems, and image processing. Two common types of digital filters are Finite Impulse Response (FIR) filters and Infinite Impulse Response (IIR) filters.

6.3.1   FIR FILTERS

FIR filters are characterized by having a finite impulse response, meaning that their output response to an impulse function input lasts for a finite duration.

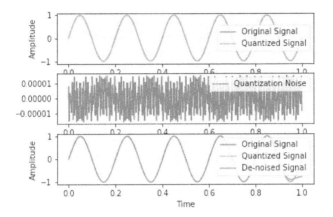

Figure 6.1   Denoising of Quantization Noise Using LPF

Mathematically, the output of an FIR filter is a weighted sum of the past and present input samples. The merits of FIR filters are:

1. *Stability:* FIR filters are inherently stable, as they do not rely on feedback loops.
2. *Linear Phase Response:* FIR filters can achieve a linear phase response, ensuring that all frequencies experience the same delay, making them suitable for applications where phase distortion needs to be minimized.
3. *Design Flexibility:* Designing FIR filters allows for precise control over the filter characteristics, including the ability to easily design filters with desired frequency response characteristics.

The demerits of FIR filters are:

1. *High Computational Complexity:* FIR filters often require more computational resources compared to IIR filters, especially for high-order designs.
2. *Delay:* Due to their finite impulse response, FIR filters introduce a delay in the signal, which might be a concern in real-time applications.

6.3.2   IIR FILTERS

IIR filters have an infinite impulse response, meaning that the output response to an impulse input continues indefinitely. IIR filters employ feedback loops in their design. The merits of IIR filters are:

1. *Lower Computational Complexity:* IIR filters generally require fewer coefficients and computations compared to FIR filters, making them computationally more efficient.
2. *Compact Design:* IIR filters can achieve similar frequency response characteristics with fewer parameters, resulting in a more compact design.

The demerits of IIR filters are:

1. *Potential Stability Issues:* IIR filters can be prone to stability issues, especially if not designed carefully, leading to problems like limit cycles or overflow.
2. *Non-Linear Phase Response:* IIR filters may introduce non-linear phase responses, which could be a concern in applications where phase linearity is crucial.

In summary, FIR and IIR filters have their strengths and weaknesses, and the choice between them depends on the specific requirements of the application. FIR filters are often preferred when stability and linear phase response are critical, while IIR filters may be chosen for applications with limited computational resources and when a more compact design is essential.

6.4 DESIGN OF DIGITAL LOW PASS FILTER

The simulation of digital Low Pass Filter in MATLAB, Octave, Mathematica, and Python is quite similar to that of an analog filter. Depending on the frequency response characteristics needed for a specific application, which includes the passband ripple, the width of the transition band, and the stopband attenuation, we can design and simulate filters with different response types. These include *Butterworth, Chebyshev, Bessel, and Elliptic filters*. Each one of them is characterized by a typical response characteristics. For example, a low pass filter of Butterworth type shows an almost flat passband response (in other words, it has minimal passband ripple); and a wider transition band, and stopband with moderate attenuation. In the following subsection, we will discuss a typical digital low pass filter design using the MATLAB Signal Processing Toolbox.

6.4.1 DIGITAL FILTER DESIGN AND SIMULATION USING DSP SYSTEM TOOLBOX

The Signal Processing Toolbox/Signal Library in SciPy Python contains functions to design and analyze digital filters, including Butterworth, Chebyshev Type I and II, Bessel, and elliptic. It is also possible to perform analog-to-digital filter conversion using discretization methods such as impulse invariance and the bilinear transformation. Once a filter is designed, the filtering process can be simulated using the *filtfilt()* available in MATLAB and SciPy.

6.4.1.1 Design of a Typical Digital Butterworth LPF

The MATLAB/GNU Octave function buttord(.) computes the order of the Butterworth filter for a given set of parameters like passband-edge frequency, stopband-edge frequency, allowable passband ripple in dBs, and the minimum attenuation in the stopband in dBs. We can combine the buttord(.) function with buttap(.) to design analog filters of Butterworth type, given appropriate specifications as mentioned above. For example, the command $[N, Wn] = buttord(Wp, Ws, Rp, Rs)$ returns the order N of the lowest order Butterworth filter which has a passband ripple of no more than *Rp* dB and a stopband attenuation of at least *Rs* dB. *Wp* and *Ws* are the passband and stopband edge frequencies, normalized between $(0, 1)$; where 1 corresponds to π radians/sample. Also, for a lowpass filter, we need to ensure that $Ws > Wp$.

Similarly, for designing digital lowpass filters of other genre like Chebyshev Type I and II, Bessel, and Elliptical there are similar function pairs like cheby1ord, cheby2ord, besselord, and ellipord. A typical MATLAB/GNU Octave script to design, simulate, and plot various types of digital low-pass filters (to compare their performance) is given below:

```
%% MATLAB Script to Design a 5th-order digital Butterworth lowpass
% filter with a cutoff frequency (f) of 4MHz. Divide f  by fs/2 to
% nomalize  the frequency to be in (0, 1). Compute the frequency
% response of the filter at 4096 points & Compare with other types.
n = 5; % Order of the Butterworth LPF = 5
f = 4e6;%Cutoff frequency = 4MHz
fs = 10e6; % Sampling frequency.
[zb,pb,kb] = butter(n,f/(fs/2)); %Normalize cutoff frequency by
%Nyquist Frequency = fs/2
[bb,ab] = zp2tf(zb,pb,kb);
[hb,wb] = freqz(bb,ab,4096);
%Design a 5th-order Chebyshev Type I filter with the same edge
% frequency  and 3 dB of passband ripple. Compute its frequency
% response.
[z1,p1,k1] = cheby1(n,3,f/(fs/2));
[b1,a1] = zp2tf(z1,p1,k1);
[h1,w1] = freqz(b1,a1,4096);
%Repeat for a 5th-order Chebyshev Type II filter with the
% same edge frequency and 30 dB of stopband attenuation
[z2,p2,k2] = cheby2(n,30,f/(fs/2));
[b2,a2] = zp2tf(z2,p2,k2);
[h2,w2] = freqz(b2,a2,4096);
% Repeat for 5th-order Elliptic filter with same parameters.
[ze,pe,ke] = ellip(n,3,30,f/(fs/2));
[be,ae] = zp2tf(ze,pe,ke);
[he,we] = freqz(be,ae,4096);
% Plot the attenuation in decibels. Express the frequency in MHz
% and  compare the performance of all four filters.
plot(wb,mag2db(abs(hb)),LineWidth=1.5)
hold on
plot(w1,mag2db(abs(h1)),LineWidth=2.0)
plot(w2,mag2db(abs(h2)),LineWidth=2.5)
plot(we,mag2db(abs(he)),LineWidth=3.0)
axis([0 6 -50 5]); grid;hold off
xlabel('Frequency  in MHz')
ylabel('Attenuation in dB')
legend('butter','cheby1','cheby2','elliptic')
```

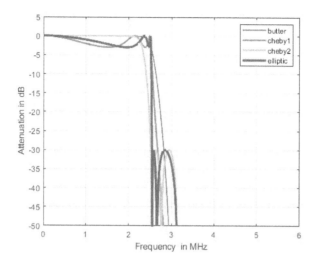

Figure 6.2   Comparison of Performances of Digital Low-Pass Filters

The output of the simulation is shown in figure 6.2.

6.5   DESIGN OF DIGITAL HIGH PASS FILTERS (HPF)

The design and simulation of digital high-pass filters using MATLAB Signal Processing Toolbox is similar to that of the low-pass filters explained in the previous section. We will develop a similar MATLAB script to compare the performance of various types of digital high pass filters of order 4 and stopband edge frequency, $4MHz$. The code is appended below and the output is shown in figure 6.3.

```
%% MATLAB Script to Design a 4th-order Digital Butterworth,
% Chebyshev Type I & II, and Elliptic highpass filters
% with a cutoff frequency of 5MHz.  Compute the frequency
% response of the filter at 4096 points.
close; clf;
n = 4; % Order of the digital HPF = 4
f = 5e6; % Cutoff frequency = 5MHz
fs = 12e6; % Sampling frequency = 12MHz
% design a digital filter parameter 'high' is used
[zb, pb, kb] = butter(n, f/(fs/2), 'high');
[bb, ab] = zp2tf(zb, pb, kb);
[hb, wb] = freqz(bb, ab, 4096);
% Design a 4th-order Chebyshev Type I filter with the same
% edge frequency  and 3 dB of passband ripple.
[z1, p1, k1] = cheby1(n, 3, f/(fs/2), 'high');
```

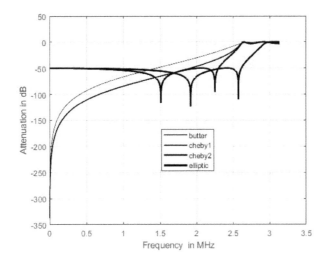

Figure 6.3 Comparison of Performances of Digital High-Pass Filters

```
[b1, a1] = zp2tf(z1, p1, k1);
[h1, w1] = freqz(b1, a1, 4096);
% Repeat for a 4th-order Chebyshev Type II filter with the
% same edge frequency and 50 dB of stopband attenuation
[z2, p2, k2] = cheby2(n,50, f/(fs/2), 'high');
[b2, a2] = zp2tf(z2, p2, k2);
[h2, w2] = freqz(b2, a2, 4096);
% Repeat for 4th-order Elliptic filter with the same parameters
[ze, pe, ke] = ellip(n, 3, 50, f/(fs/2), 'high');
[be, ae] = zp2tf(ze, pe, ke);
[he, we] = freqz(be, ae, 4096);
% Plot the attenuation in decibels. Express the frequency in MHz
% and  compare the performance of all four filters.
plot(wb, mag2db(abs(hb)), 'Color', [0, 0, 0], LineWidth=1.0)
hold on
plot(w1, mag2db(abs(h1)), 'Color', [0, 0, 0], LineWidth=1.5)
plot(w2, mag2db(abs(h2)), 'Color', [0, 0, 0], LineWidth=2.0)
plot(we, mag2db(abs(he)), 'Color', [0, 0, 0], LineWidth=2.5)
% axis([0 8 -60 5]);
grid; hold off
xlabel('Frequency  in MHz')
ylabel('Attenuation in dB')
legend({'butter', 'cheby1', 'cheby2', 'elliptic'})
```

6.6   DESIGN OF DIGITAL BAND PASS FILTERS (BPF)

In the case of band pass filters, if one examines the pass band, it is spread in between two stopbands. Band pass filters, in particular, are crucial for isolating signals within a specified frequency range. Using the library functions of MATLAB Signal Processing/DSP System Toolbox we can easily design and simulate band-pass filters of various types. We will now consider the functions available in MATLAB, Mathematica, and SciPy (Python) for the design, and simulation of digital band pass filters.

MATLAB, a widely used numerical computing environment, offers a comprehensive set of tools for designing and simulating digital bandpass filters. The Signal Processing/DSP System Toolbox in MATLAB provides functions to design various types of filters, including band-pass filters. The *designfilt* function allows users to design filters based on specifications such as passband frequency, stopband frequency, and filter order. Once the filter is designed, the *filter* function can be employed to simulate the filter's response to a given input signal. The resulting output can be visualized using the plotting capabilities of MATLAB, facilitating a thorough analysis of the filter's performance.

Mathematica, a symbolic computation software, is equipped with powerful signal-processing functions that simplify the design and simulation of digital filters. The built-in functions like *BandpassFilterModel* enable users to create a digital band pass filter based on specified parameters such as center frequency and bandwidth. Mathematica's intuitive syntax allows for a straightforward implementation of filter designs. Simulation in Mathematica for checking the performance of a filter designed, involves applying the designed filter to a signal using the *FilteringOperator* function. The software's dynamic visualization capabilities enable users to examine the time and frequency domain responses of the filter, aiding in a comprehensive analysis.

Python, with its extensive scientific computing libraries and its open-source nature, has become a popular choice for digital signal processing tasks. The *scipy* library, in particular, provides functions for designing and simulating digital filters. The *scipy.signal* module includes functions like *butter* for designing Butterworth filters, including band pass filters. Python's flexibility and readability make it easy to implement and simulate digital filters. The *scipy.signal.lfilter* function can be used to apply the designed filter to a signal, and tools like *matplotlib* can visualize the filter's response in both time and frequency domains.

A typical Python script using the SciPy.signal library suite to design and simulate digital bandpass filters of various types and to compare their performances is given below. The resulting output of the simulation is shown in figure 6.4.

```python
import matplotlib.pyplot as plt
from scipy.signal import cheby1, cheby2, ellip, freqz
import numpy as np

# Design parameters of the filter
order = 5
pb_freq = [4e6, 8e6]
ripple = 2  # Maximum passband ripple in dB
```

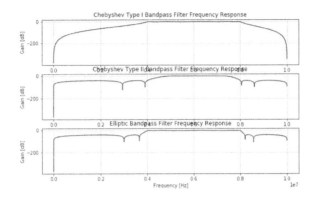

Figure 6.4   Comparison of Performances of Digital Band Pass Filters

```
sb_attn = 40  # Minimum stop band attenuation in dB
sampling_rate = 20e6  # Example sampling rate, adjust as needed
fn = 0.5 * sampling_rate
pb_freq_norm = np.array(pb_freq) / fn

# Design filters
b1, a1 = cheby1(order, ripple, pb_freq_norm, btype='band', analog=False)
b2, a2 = cheby2(order, sb_attn, pb_freq_norm, btype='band', analog=False)
b3, a3 = ellip(order, ripple, sb_attn, pb_freq_norm, btype='band', analog=False)

# Frequency response
w1, h1 = freqz(b1, a1, worN=8000)
freq_resp1 = 20 * np.log10(abs(h1))
w2, h2 = freqz(b2, a2, worN=8000)
freq_resp2 = 20 * np.log10(abs(h2))
w3, h3 = freqz(b3, a3, worN=8000)
freq_resp3 = 20 * np.log10(abs(h3))

# Plot the frequency response
plt.figure(figsize=(10,6))

plt.subplot(3, 1, 1)
plt.plot(0.5 * sampling_rate * w1 / np.pi, freq_resp1, 'b')
plt.title('Chebyshev Type I Bandpass Filter Frequency Response')
plt.xlabel('Frequency [Hz]')
plt.ylabel('Gain [dB]')
plt.grid()

plt.subplot(3, 1, 2)
plt.plot(0.5 * sampling_rate * w2 / np.pi, freq_resp2, 'b')
plt.title('Chebyshev Type II Bandpass Filter Frequency Response')
plt.xlabel('Frequency [Hz]')
plt.ylabel('Gain [dB]')
plt.grid()
```

```
plt.subplot(3, 1, 3)
plt.plot(0.5 * sampling_rate * w3 / np.pi, freq_resp3, 'b')
plt.title('Elliptic Bandpass Filter Frequency Response')
plt.xlabel('Frequency [Hz]')
plt.ylabel('Gain [dB]')
plt.grid()

# Save and show the figure
plt.savefig('fig6_4.jpg')
plt.show()
```

The following points are to be noted while designing and simulating digital band-pass and band stop filters, regarding the pass band edge frequencies, denoted by Wp radians/seconds:

1. The passband edge frequency, is specified as a two-element vector. The passband edge frequency Wp is the frequency at which the magnitude response of the filter is Rp decibels. Smaller values of passband ripple, Rp, result in wider transition bands.
2. If Wp is a scalar, then *cheby1(.)* designs a lowpass or highpass filter with edge frequency Wp.
3. If Wp is the two-element vector $[w1w2]$, where $w1 < w2$, then *cheby1* designs a bandpass or bandstop filter with lower edge frequency $w1$ and higher edge frequency $w2$.
4. For digital filters, the normalized passband edge frequencies must lie between 0 and 1, where 1 corresponds to the Nyquist rate or half the sample rate or π rad/sample.

6.6.1   THE FILTERBUILDER APP

The *filterBuilder* enables the users to design various digital filters interactively. It relies on the *fdesign* object-object oriented filter design paradigm, and is intended to reduce development time during the filter design process. filterBuilder uses a specification-centered approach to find the best algorithm for the desired response. To use the *fdesign* and *filterBuilder* it is mandatory that the Signal Processing Tool-box and the DSP System Toolbox are installed. To use filterBuilder, enter filter-Builder at the MATLAB command line using one of three approaches:

1. Enter *filterBuilder*. MATLAB opens a dialog to select a filter response type. After the user selects a filter response type, filterBuilder launches the appropriate filter design dialog box.
2. Enter *filterBuilder(h)*, where *h* is an existing filter object. For example, if *h* is a bandpass filter object, created using*filterBuilder* or *fdesign*; filterBuilder(h) opens the bandpass filter design dialog box.
3. Enter filterBuilder('response') to replace response with a response method.

See MATLAB extensive documentation on the response method, under the filter-Builder for more details.

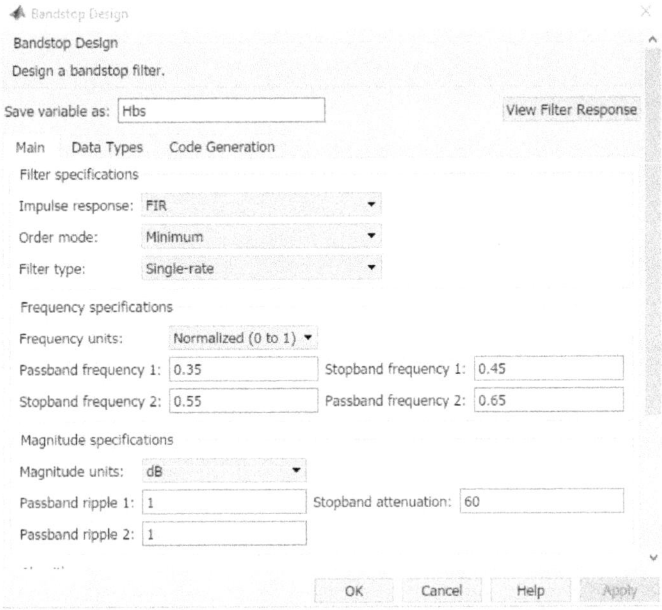

Figure 6.5   The filterBuilder Opening Window for a Band Stop Filter

To evoke the filterBuilder, type `filterBuilder('bandstop')` at the MATLAB command line. A window will open up as shown in figure 6.5. As an example, choose the following options for a digital FIR bandstop, equiripple, elliptic filter:

- Under Filter specifications, choose Impulse Response: FIR; Order Mode: Minimum; Filter Type: Single-rate.
- Under Frequency specifications, choose Frequency units: MHz; Input sample rate: 32; Passband frequency 1: 4; Stopband frequency 1: 6; Stopband frequency 2: 6; Passband frequency 2: 10.
- Under magnitude specifications, choose Magnitude units: dB; Magnitude Constraints: Constrained bands; Passband ripple 1: 1; Passband ripple 2: 1; Stopband attenuation: 60.
- Under Algorithm, choose Design method: Equiripple.
- Under Design options, Density factor: 32; Phase constraint: Linear; Wpass1: 1; Wpass2: 1.
- Under Structure select: Direct-form FIR.

After that press the Apply button at the bottom right corner. A new bandpass filter object will be created and saved with the handle *Hbs*. Now we can view the filter response by clicking the *View Filter Response* button on the top of the window. The resulting response is shown in figure 6.6.

Note that the FIR bandstop filter has a linear phase response.

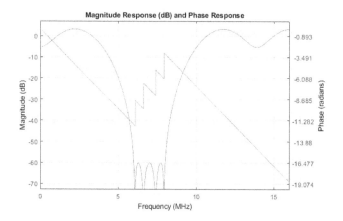

Figure 6.6 The Response of a Band Stop Filter

6.6.2 DESIGN OF DIGITAL FILTERS USING THE FDESIGN OBJECT

We can use the `fdesign` function to create a filter design specification object that contains the specifications for a filter, such as passband ripple, stopband attenuation, and filter order. Then, use the `design` function to design the filter from the filter design specifications object. Filters with different responses can be obtained by using appropriate tags to the *fdesign* function. For example, to obtain the filter objects of a digital bandstop filter, we can use the following code snippet; where all frequencies are normalized between $(0, 1)$:

```
D = fdesign.bandstop('Fp1,Fst1,Fst2,Fp2,Ap1,Ast,Ap2',.4,.5,.6,.7,1,80,.5);
% Passband edge frequency 1: 'Fp1' = 0.4;
% Stopband edge frequency 1: 'Fst1' = 0.5;
% Stopband edge frequency 2: 'Fst1' = 0.6;
% Passband edge frequency 2: 'Fp2' = 0.7;
% Ripple in Passband 1: 'Ap1' =1dB;
% Attenuation in stop band: 'Ast' = 80dB;
% Ripple in Passband 2: 'Ap2'=0.5dB;
H = design(D,'equiripple'); % design a digital filter object
freqz(H);% Plot the frequency and phase response of the digital filter.
```

Once the above code is executed the *fvtool* window will open up automatically. The resulting magnitude and phase responses are shown in figure 6.7.

We can obtain other response types by evoking the *fdesign()* with appropriate tags, as listed below:

- fdesign.lowpass; constructs a lowpass filter.
- fdesign.highpass; constructs a highpass filter.
- fdesign.decimator; constructs a decimator filter.
- fdesign.bandpass; constructs a bandpass filter.

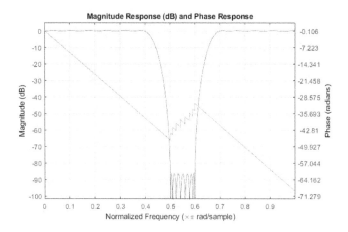

Figure 6.7   The Response of a Band Stop Filter Using fdesign(.) Function

- fdesign.differentiator; creates a default differentiator filter designer D with the filter order set to 31.
- fdesign.comb; creates a comb filter.
- fdesign.hilbert; creates an object to specify an FIR Hilbert transformer.

Other specifications can be obtained from the MATLAB documentation.

6.7   DESIGN AND SIMULATION OF DIGITAL COMB FILTERS

Digital comb filters play a crucial role in various signal processing applications, offering effective solutions for tasks such as noise reduction, interference cancellation, and signal enhancement. MATLAB provides an ideal platform for designing and simulating digital comb filters due to its extensive toolboxes and user-friendly interface. A digital comb filter is characterized by its periodic frequency response, resembling the teeth of a comb. It is widely employed in digital signal processing to selectively pass or reject specific frequency components within a signal. To design a comb filter with a *notch* or *peak* characteristics; we can either use *fdesign.comb('notch', 'N,BW',5,0.2)* or *fdesign.comb('peak', 'N,BW',5,0.2)* respectively. Here 'N' is the order of the filter = 5; and BW = 0.2dB is the bandwidth of the notch or peak. See MATLAB documentation for detailed information. A typical MATLAB code snippet to design, simulate digital comb filter is given below:

```
%% MATLAB Script to design & visualize digital Notch
%  Comb Filter.
% Comb filter design parameters
N = 6;  % Order of the Comb filter
BW= 0.2;  % Bandwidth of the comb filter
d  = fdesign.comb('notch','N,BW',N,BW);
```

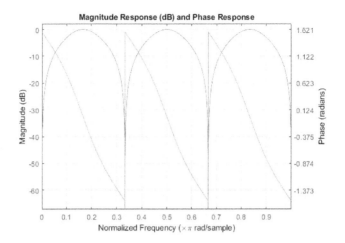

Figure 6.8   Response of a Digital Comb Notch Filter

```
Hd = design(d,'SystemObject',true);
fvtool(Hd);
% Frequency & Phase response visualization.
```

The response of the Comb filter is shown in figure 6.8

6.8   DIGITAL ALL-PASS FILTERS

All-pass IIR filters are essential components in signal processing, designed to allow all frequencies to pass through while altering the phase relationship between them. Unlike traditional filters that attenuate certain frequencies, all-pass filters retain the amplitude but modify the phase. This unique characteristic finds applications in audio systems, where precise phase adjustments are crucial for achieving desirable sound characteristics. All-pass filters are employed in equalization, phase correction, and creating audio effects. They contribute to the nuanced shaping of audio signals, enhancing the overall quality and spatial perception in audio reproduction systems, making them a fundamental tool in the realm of audio signal processing. All Pass Filters have transfer function coefficients with the *reflective structure*:

$$H_{all}(z) = \frac{a_n + a_{n-1}z^{-1} + \ldots + z^{-n}}{1 + a_1 z^{-1} + \ldots + a_n z^{-n}} = \frac{z^{-n}a(z)}{a(z^{-1})} \qquad (6.8.1)$$

Notice that the numerator polynomial is just the same as the denominator polynomial $a(z^{-1})$, but with the coefficients reversed. A sample MATLAB script to simulate the amplitude and phase response of a 10th order All-Pass Filter is given below. The output of the simulation is shown in figure 6.9.

```
%% MATLAB script to design & plot
% the response of a 10th order All Pass filter.
```

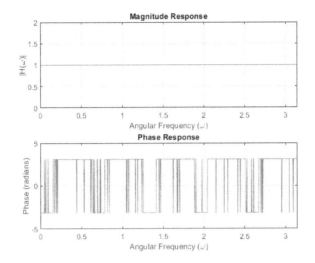

Figure 6.9   Response of a Digital All-Pass Filter

```
% Order of the all-pass filter
N=10; % order = 10;

% Angular frequency range
omega = linspace(0, pi, 1000);

% Design the all-pass filter coefficients
b = [1 zeros(1, N-1) -1];
a = [-1 zeros(1, N-1) 1];

% Compute the frequency response
H = freqz(b,a,omega);

% Plot the magnitude and phase response.

% Magnitude response
subplot(2,1,1);
plot(omega, abs(H));
title('Magnitude Response');
xlabel('Angular Frequency (\omega)');
ylabel('|H(\omega)|');
axis([0 pi 0 2]); grid;

% Phase response
subplot(2,1,2);
```

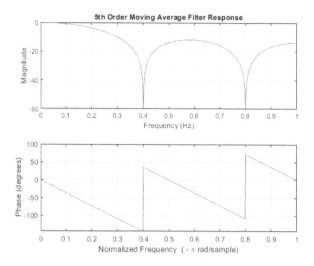

Figure 6.10   Response of a Digital Moving Average Filter

```
plot(omega, angle(H));
title('Phase Response');
xlabel('Angular Frequency (\omega)');
ylabel('Phase (radians)');
axis([0 pi -5 5]);grid;
```

6.9   DESIGN OF MOVING AVERAGE FILTER

A digital moving average filter is employed in signal processing to smooth out variations or noise in a time-series data set. It calculates the average of a specified number
of consecutive data points, creating a moving average. This filtering technique is
useful for eliminating short-term fluctuations or high-frequency noise, revealing the
underlying trends or patterns in the signal. Digital moving average filters find applications in diverse fields, such as finance, audio processing, and sensor data analysis,
where it's crucial to enhance data quality by reducing random variations and obtaining a clearer representation of the underlying signal dynamics. MATLAB/GNU
Octave script to design and plot a 5th-order moving average filter is given below. The
output of the simulation is shown in figure 6.10.

```
%% MATLAB script to design and plot
% 5th Order Digital Moving Average Filter.
% Order of the moving average filter
order = 5;
```

```
% Generate the filter coefficients
b = ones(1, order)/order;% numerator polynomial coefficients
a =1; % denominator polynomial coefficient
% Frequency response of the filter
freqz(b, a);

% Plot the magnitude and phase responses
title('5th Order Moving Average Filter Response');
xlabel('Frequency (Hz)');
ylabel('Magnitude');
```

6.10   WAVELET DENOISING AND WIENER FILTERING

Wavelet denoising uses wavelet transforms to remove noise from signals while preserving their essential features. It decomposes the signal into different frequency components, allowing the targeted noise removal. Wiener filtering is a statistical signal processing method that minimizes mean-squared error between the original signal and the estimated signal. Applied in the frequency domain, Wiener filtering is effective for reducing additive noise. Both techniques are widely used in various fields, including image processing and audio signal enhancement, to improve the quality and clarity of signals in the presence of noise.

6.10.1   WAVELET DENOISING

Wavelet denoising using Daubechies 2 wavelet is discussed in this section. A sample MATLAB code to denoise and plot the noisy signal is given below. The output of the simulation is shown in figure 6.11.

```
%% MATLAB script to demonstrate Wavelet denoising.
% Generate a signal with noise
t = linspace(0, 1, 1000);
signal = sin(2 * pi * 5 * t) + 0.5 * randn(size(t));

% Plot the original signal with noise
subplot(2, 1, 1);
plot(t, signal);grid; axis([0 1 -3 3])
title('Original Signal with Noise');
xlabel('Time'); ylabel('Amplitude');

% Apply wavelet denoising
level =2; % threshold level=0.2;
% Adjust this threshold based on the
% signal characteristics
denoised_signal = wdenoise(signal,level,'Wavelet', 'db1');
```

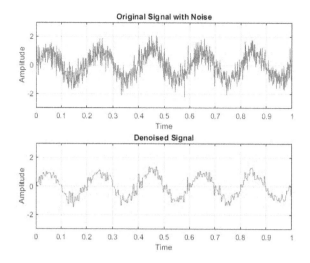

Figure 6.11   Response of a DB2 Wavelet Filter

```
% Plot the denoised signal
subplot(2, 1, 2);
plot(t, denoised_signal);
title('Denoised Signal');
grid; axis([0 1 -3 3]);
xlabel('Time'); ylabel('Amplitude');
```

6.10.2   WIENER FILTERING

Wiener filtering is a signal processing technique that minimizes mean-squared error between the estimated and true signals. It is effective for noise reduction in signals. In MATLAB, the `wiener2(.)` function implements Wiener filtering. A MATLAB code snippet is given below. The simulated output is shown in figure 6.12.

```
%% MATLAB script to demonstrate Wiener filtering.
clear; close;clf;
noisy_signal = sin(2*pi*0.05*(1:500)) + 0.2*randn(1,500);
filtered_sig = wiener2(noisy_signal);
plot(noisy_signal, 'b','DisplayName','Noisy Signal');
hold on; grid;xlabel('Time');ylabel('Amplitude')
plot(filtered_sig,'r','DisplayName','Wiener Filtered Signal');
legend;
```

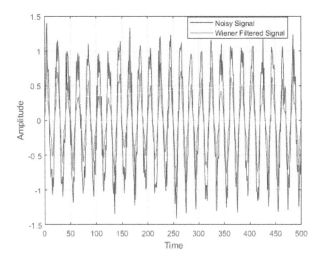

Figure 6.12 Response of a Wiener Filter

6.11 CONCLUDING REMARKS

Digital signal processing plays a pivotal role in various applications, ranging from communication systems to biomedical devices. Among the essential components of digital signal processing are filters, which help in extracting or modifying specific frequency components of a signal. This chapter focused on the design and simulation of digital band pass filters using two powerful programming environments, namely MATLAB and Python.

In this chapter, we also discussed the design and simulation of digital low pass, high pass, band pass, band stop, comb filter, moving average filter, Wavelet filter and Wiener Filter using MATLAB. All the relevant MATLAB library functions in Signal Processing, DSP System, and Communication Toolboxes were also introduced.

In conclusion, the design and simulation of digital filters are foundational aspects of digital signal processing, playing a critical role in shaping the performance of various systems across diverse domains. This chapter has provided a comprehensive exploration of digital filters, showcasing the versatility of MATLAB and Python with the SciPy library.

MATLAB, with its rich set of tools and functions, offers an intuitive environment for filter design and simulation. Its extensive signal processing toolbox, coupled with powerful plotting capabilities, empowers engineers and researchers to effortlessly prototype and analyze digital filters. MATLAB's user-friendly interface facilitates quick experimentation and iteration, making it a preferred choice in academic, research, and industrial applications.

On the other hand, Python, with the SciPy library, emerges as a flexible and open-source alternative. The script-based nature of Python, coupled with the modular structure of SciPy, allows for seamless integration into larger workflows. This combination is particularly advantageous in applications where customization, automation, and scalability are paramount. As technology continues to advance, the demand for efficient and adaptive digital filtering solutions grows. Whether in communications, biomedical applications, or audio signal processing, the knowledge gained from this chapter equips practitioners to navigate the complexities of filter design and simulation.

FURTHER READING

1. Luis F. Chaparro, *Signals and Systems Using MATLAB*, Cengage Learning.
2. Robert J. Schilling and Sandra L. Harris, *Introduction to Digital Signal Processing Using MATLAB*, 2nd Edition.
3. B.P. Lathi, *Signal Processing & Linear Systems*, Oxford University Press, 2008.
4. The MathWorks Inc., *DSP System Toolbox Reference*, 2023.
5. The MathWorks Inc., *DSP System Toolbox User Guide*, 2023.
6. The MathWorks Inc., *Communications Toolbox Reference*, 2023.
7. The MathWorks Inc., *Signal Processing Toolbox User Guide*, 2023

EXERCISES

1. Design, simulate, and plot the magnitude and phase response of a Butterworth digital low pass filter with the following specifications: cut-off frequency=60Hz, maximum pass band ripple = 0.1dB, and minimum stop band attenuation=40dB. Choose a suitable sampling frequency. Decide the order of the filter using the MATLAB function, *buttord(.)* and proceed. What is your observation?
2. Repeat exercise 1 with the following modifications: design a Chebychev Type I digital high pass filter, with passband edge frequency $2kHz$, pass band ripple=2dB, and stop band attenuation=60dB.
3. Write MATLAB/GNU Octave script to design, simulate, and plot the responses of a Butterworth, Chebyshev Type I & II, and Elliptic digital low pass filter with the following specifications: cut-off frequency=60kHz; maximum pass band ripple = 1dB; minimum stop band attenuation = 60dB.
4. Design, simulate, and plot the amplitude and frequency response of an digital high pass filter with following specifications: pass band edge frequency=2MHz; maximum pass band ripple=0.2dB; minimum stop band attenuation=60dB. Filter type: elliptic.
5. Design, simulate, and plot the response of a digital band pass filter with following specification using MATLAB/GNU Octave: pass band ripple = 1dB; pass band edge frequencies: 6MHz and 10MHz; minimum attenuation in stop band = 60dB; Filter type: Butterworth, Chebyshev Type II. What is the inference?
6. Assume that an Additive White Gaussian Noise (AWGN) having variance of 0.1 is added to a typical sinusoidal signal of amplitude 2V-pp and frequency 1kHz. De-noise the above signal using a suitable digital low pass filter Develop suitable MATLAB/GNU Octave script for the above.

7. Design, simulate, and plot the response of a digital band stop filter with following specification using MATLAB/GNU Octave: pass band ripple = 2dB; pass band edge frequencies: 8GHz and 12GHz; minimum attenuation in stop band = 60dB; Filter type: Chebyshev Type I and II; and Elliptic. What is the inference?

8. Design, simulate, and plot the response of a digital band pass filter with following specification using MATLAB/GNU Octave: pass band ripple = 1dB; pass band edge frequencies:10MHz and 16MHz; minimum attenuation in stop band = 60dB; Filter type: Chebyshev Type I and II; and Elliptic.

9. Design, simulate, and plot a digital comb filter with following parameters: order N=30; response type = 'notch'; Number of peaks =5; Band Width = 20; Gain at bandwidth=-6dB, Shelving filter order=6, and frequency=600Hz.

10. Design, simulate, and plot a digital comb filter with following parameters: order N=25; response type = 'peak'; Number of peaks =5; Band Width = 10; Gain at bandwidth=-6dB, Shelving filter order=5, and frequency=800Hz.

7 Introduction to Modern Communication Systems

In this chapter, we discuss the design and simulation of some of the modern communication systems, viz. MIMO using the Signal Processing, Antennas, Communications, and DSP System Toolboxes of MATLAB. Multiple Input Multiple Output (MIMO) systems are a recent promising technological trend in the annals of communication systems engineering. The MIMO technology is widely used in smart cellular phones (third generation and beyond-mobile phones) so that signal reception is hassle-free and of desired quality. With the implementation of MIMO technology the signal quality in cell phones has improved considerably. This chapter aims to explore the evolution of MIMO systems, from their conceptualization to current applications, and provide practical insights into simulating MIMO systems using MATLAB and Python.

7.1 INTRODUCTION

In recent years, it was realized that the MIMO communication systems seem to be inevitable in accelerated evolution of high data rate applications due to their potential to dramatically increase the spectral efficiency and simultaneously sending individual information to the corresponding users in wireless systems. Multiple Input Multiple Output (MIMO) communication systems have become a cornerstone in modern wireless communication, providing significant advancements in terms of data rate, reliability, and spectral efficiency. The concept of MIMO dates back to the early 1970s, with initial research focusing on simple point-to-point communication. Over the years, MIMO technology has evolved from its theoretical foundations to practical implementations. The breakthrough came with the advent of space-time coding and spatial multiplexing techniques.

In major cellular and wireless networks today, space diversity is employed with the help of multiple transmit antennas and/or multiple receive antennas giving rise to Multiple Input Multiple Output (MIMO) systems. There are three different modes in which multiple antennas can be deployed:

- Beamforming
- Spatial Multiplexing
- Space-Time Coding

7.1.1 BEAMFORMING

It has been used by signal-processing engineers for radio applications, for quite some time. For example, Marconi used four antennas to increase the gain of signal

DOI: 10.1201/9781003527589-7

transmissions across the Atlantic, way back in 1901. It has also been known since 1970s that multiple antennas at the base station help simultaneous communication with several users. It was only after the advent of Multiple Input Multiple Output (MIMO) systems in mid 1990s, and then its adoption as the defining technology for 5G mobile networks, that beamforming became a household terminology.

In MIMO systems, where multiple antennas are employed at both the transmitter and receiver ends, beamforming optimizes the use of these antennas to improve data rates, reliability, and coverage. Beamforming in MIMO systems is a sophisticated signal processing technique that enhances wireless communication performance by directing radio frequency signals toward specific spatial directions. In MIMO setups, where multiple antennas are employed at both the transmitter and receiver ends, beamforming optimizes the use of these antennas to improve data rates, reliability, and coverage. Traditional MIMO systems exploit spatial diversity to transmit multiple data streams simultaneously, but beamforming takes it a step further. It focuses radio frequency energy in specific directions, forming a "beam" that maximizes signal strength and quality at the intended receiver, while minimizing interference and signal degradation in other directions. This targeted approach significantly increases the system's overall capacity and spectral efficiency.

There are two main types of beamforming in MIMO systems: transmit beamforming and receive beamforming. Transmit beamforming is implemented at the transmitter to enhance signal strength at the receiver, while receive beamforming is applied at the receiver to improve the reception of signals from a specific direction. By dynamically adjusting the phase and amplitude of signals across multiple antennas, MIMO beamforming optimizes the spatial dimension of wireless communication, leading to improved data rates, reduced interference, and enhanced overall system performance. This technology is crucial for meeting the increasing demand for high-capacity and reliable wireless communication in modern networks.

The confusing aspect of beamforming arises due to its inherent association with the idea of directionality that does not make sense in urban cellular networks (with buildings, trees, vehicles, and other objects) posing an obstacle to the Line of Sight (LoS) paths. It is an antenna array that 'looks' only into a particular intended direction thus minimizing the interference for other users in the same area. This is the essence of classical beamforming which is implemented through varying the phase and/or amplitude of the signal at each antenna; which is similar to burning a paper with a magnifying glass under the sun.

7.1.2   SPATIAL MULTIPLEXING

Spatial multiplexing in wireless communications is based on Multiple Input Multiple Output (MIMO) technology where multiple antennas at both the transmitter and receiver are used to carry multiple data streams simultaneously within the same frequency band. There is a subtle difference between beamforming and spatial multiplexing. Generalized beamforming makes use of only one of the modes available, while Spatial Multiplexing makes use of all of them. Low complexity is one of the main advantages of the beamforming approach. However, beamforming may be suboptimal. Spatial multiplexing has the ability to approach the maximum channel capacity. Spatial diversity is a technique in MIMO that reduces signal fading by sending multiple copies of the same radio signal through multiple antennas; and there is a difference between Spatial Diversity and Spatial Multiplexing. Spatial multiplexing boosts data rates by sending the data payload in separate streams through spatially separated antennas. The following are the main points to be noted with respect to spatial multiplexing:

- It does not require bandwidth expansion.
- It needs space-time equalization at the receiver.
- Data streams can be separated by the equalizers.
- It is a better alternative to spatial diversity.
- Independent spatial channels give accurate results in transmission.

In short, spatial multiplexing focuses on transmitting multiple independent data streams simultaneously to increase data throughput, while spatial diversity focuses on transmitting redundant copies of the same data stream to improve reliability and combat fading effects. In practice, the MIMO systems often utilize a combination of these techniques, depending on the communication application and objectives.

7.1.3   SPACE-TIME CODING

Space-time coding achieves transmit diversity through multiple antennas at the transmitter side and simple linear processing at the receiver. This simplicity made this technique quite suitable for the past generations of cellular and other infrastructure based networks. There are two main kinds of space-time codes: Space-Time Block Codes (STBC) and Space-Time Trellis Codes (STTC). In general, STTC offer a better performance than STBC at a cost of increased computational complexity.

Space-time coding involves encoding data in both the spatial and temporal domains, exploiting the diversity provided by multiple antennas. This technique helps combat fading and improves the robustness of the transmitted signal. STC employs complex algebraic manipulations to create a matrix of symbols, optimizing transmission for various paths between antennas.

On the other hand, Space-Time Block Codes specifically focus on transmitting data over multiple antennas in a block-wise fashion. It arranges data into blocks and encodes them across both space and time dimensions, ensuring better error resilience and improved diversity gain. Space-Time Block Codes is known for its simplicity in implementation and compatibility with various modulation schemes.

Both space-time coding and space-time block coding play pivotal roles in achieving reliable and high-capacity wireless communication. By capitalizing on the spatial dimension, these techniques mitigate the challenges posed by channel fading and interference, making them indispensable in modern communication systems. Their application extends to various wireless communication standards, including WiFi, 4G, and 5G, contributing significantly to the efficiency and performance of MIMO-enabled communication networks.

7.2 MULTIPLE INPUT MULTIPLE OUTPUT (MIMO) SYSTEMS

Let us now look into some of the key technology enablers that lead to the development of MIMO systems.

- *Space-Time Coding:* Introduced in the 1990s, space-time codes like Alamouti codes played a crucial role in improving MIMO system reliability and performance by addressing fading channel issues. Space-Time Coding (STC) is a crucial technique employed in Multiple-Input Multiple-Output (MIMO) communication systems to enhance reliability and combat fading effects. It utilizes multiple antennas at both the transmitter and receiver ends to improve the robustness of the communication link. The primary goal of STC is *error reduction*; the errors are due to channel impairments such as fading and noise. The other goal is the *diversity gain* achieved by exploiting spatial diversity, which enhances system reliability in challenging propagation environments.
- *Spatial Multiplexing:* Leveraging multiple antennas for simultaneous data transmission, spatial multiplexing significantly increased data rates, paving the way for high-throughput communication.

The other key developments that lead to the evolution of MIMO systems and its widespread adoption are the following:

1. *Advancements in Antenna Technology:* Multiple antennas at both the transmitter and receiver ends are essential for MIMO systems. The development of compact, cost-effective, and high-performance antennas has been a key enabler.
2. *Signal Processing Techniques:* Signal processing algorithms, such as spatial multiplexing and beamforming, have been developed to maximize the data rate and improve the reliability of wireless communications in MIMO systems.
3. *Smart Antenna Systems:* Smart antenna systems, which can dynamically adjust the radiation pattern and directionality of antennas, are critical for achieving spatial diversity and spatial multiplexing gains in MIMO systems.
4. *Channel State Information (CSI) Feedback:* Accurate knowledge of the channel state information is crucial for the success of MIMO systems. Techniques for estimating and feeding back channel information, such as precoding matrices, have been developed, that aided the use of MIMO systems.
5. *Orthogonal Frequency Division Multiplexing (OFDM):* OFDM is a modulation technique that efficiently divides the available spectrum into multiple orthogonal subcarriers. MIMO systems often use OFDM to cope with frequency-selective fading and to enhance spectral efficiency.
6. *Increased Processing Power:* The availability of powerful and cost-effective signal processing hardware, including digital signal processors (DSPs) and application-specific integrated circuits (ASICs), has facilitated the implementation of complex MIMO algorithms.
7. *Improved Error Correction Coding:* Advanced error correction coding schemes, such as space-time block codes and spatial multiplexing, have been developed to

enhance the reliability of MIMO systems by mitigating the effects of fading and interference.

8. *Multiple Antenna Techniques (Spatial Diversity and Spatial Multiplexing):* Spatial diversity and spatial multiplexing are fundamental techniques in MIMO systems. Spatial diversity utilizes multiple antennas to combat fading, while spatial multiplexing enables simultaneous transmission of multiple data streams.

9. *Advanced MIMO Standards:* Integration of MIMO into wireless communication standards, such as WiFi (e.g., 802.11n/ac/ax) and LTE/5G, has driven the widespread adoption of MIMO technologies in consumer devices and cellular networks.

10. *Research and Standardization Efforts:* Continuous research, development, and standardization efforts by academia and industry have played a significant role in advancing MIMO technologies and ensuring interoperability across different systems.

These technology enablers collectively contributed to the success of MIMO systems, allowing them to provide increased data rates, improved link reliability, and enhanced spectral efficiency in wireless communication networks.

7.2.1   SPACE-TIME CODING (STC)

Space-time coding involves transmitting data not only in different spatial dimensions (multiple antennas) but also over multiple time instances. This combination helps combat fading and improves the overall link quality. Space-Time Coding uses *Alamouti Codes*, which is one of the earliest and widely adopted space-time coding schemes, introduced by Siavash M. Alamouti. It is particularly popular due to its simplicity and effectiveness in providing diversity gain.

The Alamouti coding scheme employs a 2×2 matrix representation for the transmitted signal. In a two-antenna system, the transmitted signal is a linear combination of the symbols from the two antennas, arranged in a specific matrix structure. The second key feature of Space-Time Coding is the orthogonality between the signals transmitted from different antennas. This orthogonality enables the receiver to separate the signals accurately, even in the presence of fading and noise.

7.2.1.1   The Working of the Alamouti Coding Scheme:

The following is the modality of operation of Alamouti coding:

- In transmitting phase 1, the data symbols from the first antenna are transmitted during the first time slot.
- In transmitting Phase 2, the complex conjugates of the data symbols from the second antenna are transmitted during the second time slot, with a sign inversion.

This process is then repeated for subsequent time slots, providing diversity at the receiver. Note that the second antenna transmits nothing (i.e. it is silent) during the first time interval; while the first antenna is silent during the second time interval. These gaps in transmissions can be efficiently utilized to build a code as follows. Benefits of Alamouti Coding scheme include *Diversity Gain* achieved by transmitting the same information through different channels simultaneously. The other merit is simplicity, which makes Alamouti coding attractive for practical implementations, especially in cases where complexity needs to be minimized. The Alamouti matrix is a key component of this coding scheme. The Alamouti matrix for two transmit antennas over two-time slots (symbols) can be represented as follows:

$$AM_{2\times2} = \begin{bmatrix} s_1 & s_2 \\ -s_2^* & s_1^* \end{bmatrix}.$$

Here, s_1 and s_2 are the symbols to be transmitted in the first and second-time slots, respectively; and $*$ denotes complex conjugation. For a more general MIMO case with N transmit antennas and N time slots, the Alamouti matrix can be extended accordingly. In a 4×4 MIMO system, the Alamouti matrix can be constructed for two consecutive time slots. The Alamouti matrix for this case is as follows:

$$AM_{4\times4} = \begin{bmatrix} s_1 & s_2 & s_3 & s_4 \\ -s_2^* & s_1^* & -s_4^* & s_3^* \\ -s_3 & -s_4 & s_1 & s_2 \\ s_4^* & -s_3^* & -s_2^* & s_1^* \end{bmatrix}.$$

Typical MATLAB and Python code snippets for a 2×2 MIMO system using Alamouti encoding is given below:

```
%% MATLAB/GNU Octave Function for
% Encoding a symbol set using the Alamouti encoder..
% Alamouti encoding function
function encoded_symbols = alamouti_encode(symbols)
    N = length(symbols);
    encoded_symbols = zeros(2, N);
    for i = 1:2:N
        % Alamouti matrix for two time slots
        Alamouti_matrix = [symbols(i), symbols(i+1); ...
        -conj(symbols(i+1)), conj(symbols(i))];
        encoded_symbols(:, i:i+1) = Alamouti_matrix;
    end
end
% Sample usage of Alamouti Encoder
symbols_to_transmit = randn(1, 4) + 1i * randn(1, 4);
% Replace with symbols to be transmitted
encoded_symbols = alamouti_encode(symbols_to_transmit);
disp('Symbols_to_Transmit:');
disp(symbols_to_transmit);
disp('Encoded symbols:');
disp(encoded_symbols);
```

An equivalent Python code using the numpy library is given below:

```python
import numpy as np
# Alamouti encoding function
def alamouti_encode(symbols):
    N = len(symbols)
    encoded_symbols = np.zeros((2, N), dtype=complex)

    for i in range(0, N, 2):
        # Alamouti matrix for two time slots
        Alamouti_matrix = np.array([[symbols[i], symbols[i+1]],
            [-np.conj(symbols[i+1]), np.conj(symbols[i])]])
        encoded_symbols[:, i:i+2] = Alamouti_matrix

    return encoded_symbols
# Example usage
symbols_to_transmit = np.random.randn(4) + 1j * np.random.randn(4)
 # Replace with actual symbols
encoded_symbols = alamouti_encode(symbols_to_transmit)
print('Encoded symbols:')
print(encoded_symbols)
```

The channel models used with MIMO systems modeling are *Rayleigh Fading Channel*, which the effects of multipath propagation and fading; as well as *Additive White Gaussian Noise (AWGN)* that represents the background noise in the communication channel.

7.2.1.2 MATLAB Script for Demonstrating MIMO Systems

A typical MATLAB code snippet is given below to demonstrate the working of a MIMO system that uses a 4QAM Modulation scheme. The output of the simulation is also given at the end of the code snippet. Note that this is a mere demonstration of the working of MIMO systems, and the value obtained for the bit error rate (BER) is not indicative of a practical system.

```matlab
%% MATLAB Script to demonstrate MIMO system.
% Define MIMO system parameters
clear;
numTxAn = 2; % Number of antennas at the transmitter
numRxAn = 2; % Number of antennas at the receiver
snr_dB = 30; % Signal-to-noise ratio in dB=30
% Generate random 5000 data symbols in [0,1].
dataSyms = randi([0, 1], 1, 5000);
% Modulation 4QAM
modSymbls = qammod(dataSyms, 4);
% Reshape data for transmission
txData = reshape(modSymbls, numTxAn, []);
```

```
% MIMO channel matrix (Rayleigh fading)
H = (randn(numRxAn,numTxAn)+1i*randn(numRxAn,numTxAn))/sqrt(2);
% Transmit through MIMO channel
rxData = H * txData;
% Add AWGN
noisePower = 10^(-snr_dB / 10);
rxDataNois = awgn(rxData, snr_dB,'measured');
% Perform MIMO decoding
decSymbls = qamdemod(H'*rxDataNois, 4, 'OutputType', 'bit');
% For consistency, reshape decodedSymbols to a row vector
decSymbls = reshape(decSymbls, 1, []);
% Calculate bit error rate
L=numel(dataSyms); % Number of dataSymbols.
ber = sum(decSymbls(1:L)~=dataSyms)/L;
disp(['Bit Error Rate: ', num2str(ber)]);
>> mimodem % Name of the script to run..
Bit Error Rate: 0.479
```

7.2.2 SPACE-TIME BLOCK CODES (STBC)

The Space-Time Block Coding is a technique used in wireless communications to transmit multiple copies of a data stream across a number of antennas and to exploit the various received versions of the data to improve the reliability of data transfer. The fact that the transmitted signal must traverse a potentially difficult environment with scattering, reflection, refraction, and so on and may then be further corrupted by thermal noise in the receiver means that some of the received copies of the data may be closer to the original signal than others. This redundancy results in a higher chance of being able to use one or more of the received copies to correctly decode the received signal. In fact, space-time coding combines all the copies of the received signal in an optimal way to extract as much information from each of them as possible.

STBC is a new paradigm for communication over Rayleigh fading channels using multiple transmit antennas. Data is encoded using a space-time block code and the encoded data is split into n streams which are simultaneously transmitted using n transmit antennas. The received signal at each receive antenna is a linear superposition of the n-transmitted signals perturbed by noise. Maximum-likelihood decoding is achieved in a simple way through decoupling of the signals transmitted from different antennas rather than joint detection. This uses the orthogonal structure of the space-time block code and gives a maximum-likelihood decoding algorithm which is based only on linear processing at the receiver. Space-time block codes are designed to achieve the maximum diversity order for a given number of transmit and receive antennas subject to the constraint of having a simple decoding algorithm. The classical mathematical framework of orthogonal designs was applied to construct space-time block codes. It was shown that space-time block codes constructed in this way only exist for few sporadic values of n. Subsequently, a generalization of orthogonal designs is shown to provide space-time block codes for both real and complex constellations for any number of transmit antennas. As reported in Ref.1, these codes achieved the maximum possible transmission rate for any number of transmit antennas using any arbitrary real constellation such as PAM. For an arbitrary complex constellation such as PSK and QAM, space-time block codes, achieved 1/2 of the maximum possible transmission rate for any number of transmit antennas. For the specific cases of two, three, and four transmit antennas, space-time block codes achieved, respectively, 1/2, 2/3, and 3/4 of maximum possible transmission rate using arbitrary complex constellations. The best tradeoff between the decoding delay and the number of transmit antennas, showed that many of the codes were optimal in that sense as well.

7.2.3 SPACE-TIME TRELLIS CODES

Space-Time Trellis Codes (STTC) are a class of error-correcting codes designed for Multiple Input Multiple Output (MIMO) communication systems. STTC combines the principles of space-time coding and trellis coding to exploit spatial and temporal dimensions simultaneously, enhancing data transmission reliability in challenging wireless communication environments. STTC achieves this by creating a trellis structure that captures the spatial and temporal relationships among transmitted symbols. The decoder utilizes this structure to efficiently recover transmitted data, providing improved diversity gain and error performance.

7.3 CONFIGURATIONS OF WIRELESS MIMO SYSTEMS

As we delve deep into the three possible configurations of MIMO systems, we can find that the following options are available.

1. *Single Input Multiple Output (SIMO) System:* It is the system where a single antenna is used at the transmitter, and multiple antennas are deployed at the receiver.

2. *Multiple Input Single Output (MISO) System:* Multiple antennas are used at the transmitter, only one antenna is used at the receiver.

3. *Multiple Input Multiple Output (MIMO):* This is the MIMO implementation in its entirety. Here, multiple antennas are deployed at both the transmitter and receiver side.

The MIMO communication system has significantly higher channel capacity than the Single Input Single Output (SISO) system for the same total transmission power and bandwidth. We will now consider the Orthogonal Space Time Block Codes used in MIMO systems.

7.3.1 ORTHOGONAL SPACE-TIME BLOCK CODING (OSTBC)

Orthogonal Space-Time Block Coding (OSTBC) is a key technology in the domain of wireless communication systems, specifically designed to enhance the performance of Multiple Input Multiple Output (MIMO) systems. The fundamental principle behind OSTBC is to encode data across multiple antennas and time instances in an orthogonal manner, allowing for increased diversity gains. The key aspects of OSTBC include:

1. *Orthogonality:* OSTBC ensures that the transmitted signals across different antennas and time slots are orthogonal. This orthogonality helps in mitigating interference and combating channel fading effects, leading to improved reliability.
2. *Spatial Diversity:* By transmitting the same information across multiple antennas with suitable coding, OSTBC introduces spatial diversity. This diversity aids in overcoming the adverse effects of fading channels, providing robust communication links.
3. *Alamouti Code:* One of the most well-known OSTBC schemes is the Alamouti code, which achieves full diversity with a simple and efficient implementation. It uses a 2×2 Alamouti matrix for encoding over two-time slots, offering improved error performance.

The advantages of OSTBC include the following:

- *Increased Data Rate:* OSTBC enables higher data rates by exploiting the spatial dimension through multiple antennas. This is particularly valuable in applications where bandwidth is limited.
- *Improved Reliability:* The use of spatial diversity through OSTBC enhances the system's reliability by mitigating the impact of fading and improving the overall link quality.
- *Spectral Efficiency:* OSTBC improves spectral efficiency by allowing multiple data streams to be transmitted simultaneously, leading to better utilization of the available frequency spectrum.

The applications of OSTBCs include high-speed wireless local area networks (WLANs) and personal area networks (PANs); mobile communication systems (to combat multipath fading and enhance the link quality, contributing to improved voice and data services) and cellular networks; Broadband Wireless Access.

Orthogonal Space-Time Block Coding is a pivotal technology in the realm of wireless communication systems, offering increased data rates, improved reliability, and enhanced spectral efficiency through the use of multiple antennas and orthogonal encoding. Its integration into existing and emerging communication standards highlights its importance in shaping the future of wireless networks. As technology continues to advance, OSTBC remains a key enabler for meeting the demands of modern communication systems.

7.4   TRICKS AND TECHNIQUES IN MIMO SYSTEMS

To summarize, the following are the tricks and techniques used in implementing a 2×2 MIMO system. The application engineer or the implementor should judiciously focus on all of the below given ten aspects for a successful implementation of MIMO system.

1. *Spatial Diversity:* Leverage spatial diversity by placing antennas at different locations to mitigate fading and enhance signal robustness. Implement Alamouti coding for simple and effective space-time coding, providing diversity gains.

2. *Beamforming:* Utilize beamforming techniques to focus the transmitted signal towards the intended receiver, enhancing the signal strength. Adaptive beamforming algorithms can optimize the direction of transmission based on channel conditions.

3. *Channel State Information (CSI):* Acquire accurate and timely CSI to adapt transmission strategies based on the current channel conditions. Then, implement feedback mechanisms for the receivers to communicate channel state information, back to the transmitters.

4. *Pre-coding Techniques:* Use pre-coding schemes like Zero Forcing (ZF) or Maximum Likelihood (ML) to mitigate interference and improve signal quality at the receiver.

5. *Spatial Multiplexing:* Exploit spatial multiplexing to transmit multiple data streams simultaneously over the same frequency band, increasing the overall data rate. Implement Singular Value Decomposition (SVD) for optimal spatial multiplexing.

6. *Interference Management:* Employ interference cancellation techniques to mitigate co-channel interference and enhance system performance. Implement advanced signal processing algorithms to separate and decode overlapping signals.

7. *Diversity Combining:* Combine signals received from different antennas using techniques like maximal ratio combining (MRC) to improve reception reliability. Choose appropriate combining schemes based on the channel characteristics.

8. *Polarization Diversity:* Combine antennas with different polarization characteristics to enhance diversity and combat polarization fading effects. Utilize circular polarization for better performance in dynamic environments.

9. *Transmit Power Control:* Implement dynamic power control mechanisms to optimize the transmission power based on channel conditions, ensuring efficient use of resources.

10. *Hybrid Beamforming:* Consider hybrid beamforming architectures for a balance between performance and complexity, especially in millimeter-wave MIMO systems.

7.4.1 CONFIGURATION OF 4×4 MIMO SYSTEMS

One possible advancement of 2×2 MIMO is the 4×4 system, that is widely implemented in various wireless communication standards and technologies around the world. These implementations are found in both cellular networks and WiFi systems. Here are a few typical use cases:

- 4G Long-Term Evolution (4G LTE): Many LTE networks worldwide use 4×4 MIMO technology to enhance data rates, improve spectral efficiency, and provide better coverage. This is especially common in urban areas with high user density.

- 5G New Radio (5G NR): The 5G standard includes support for advanced MIMO configurations, including 4×4 MIMO. It is used to achieve higher data rates, increased network capacity, and improved reliability in 5G networks.
- WiFi: In the WiFi domain, the IEEE 802.11ac standard (WiFi 5) and the subsequent IEEE 802.11ax standard (WiFi 6) support 4×4 MIMO configurations. The WiFi 6, in particular, leverages 4×4 MIMO to enhance performance in high-density environments and improve overall network efficiency.
- Point-to-Point Wireless Communication: 4×4 MIMO is also commonly used in point-to-point wireless communication links, such as back-haul connections for cellular networks or in fixed wireless broadband systems. These links benefit from the increased capacity and reliability offered by 4×4 MIMO.
- Massive MIMO Deployments: Massive MIMO, which involves using a large number of antennas at both the transmitter and receiver, often incorporates4×4 MIMO configurations as part of the overall system. Massive MIMO is a key technology in the evolution of 4G and 5G networks.

These practical implementations showcase the versatility and effectiveness of 4×4 MIMO in various communication scenarios. It plays a crucial role in meeting the increasing demand for higher data rates, improved network performance, and better user experiences in modern wireless communication systems.

7.4.2 GOING BEYOND 4×4 MIMO

Going beyond 4×4 MIMO can offer several practical advantages in terms of improved system performance, increased data rates, and enhanced reliability. Some major advantages are the following:

1. *Higher Data Rates:* Increasing the number of antennas (transmit and receive) allows for the transmission of more data streams simultaneously. This results in higher data rates and increased overall system capacity.
2. *Enhanced Spatial Multiplexing:* Beyond 4×4 MIMO, systems can leverage even more spatial dimensions for multiplexing, enabling the simultaneous transmission of multiple independent data streams. This is particularly beneficial for high-capacity applications and dense user environments.
3. *Improved Coverage and Reliability:* Additional antennas can improve the coverage area and reliability of the communication system, especially in challenging environments with obstacles or signal fading. This is crucial for providing a consistent user experience across different locations.
4. *Better Channel Estimation:* With more antennas, the system gains better spatial resolution and accuracy in estimating the channel conditions. This leads to improved channel state information (CSI), enabling more effective beamforming and adaptive transmission strategies.

5. *Massive MIMO Benefits:* Going beyond 4×4 MIMO is often associated with Massive MIMO, where a large number of antennas are deployed at both the transmitter and receiver. Massive MIMO brings additional advantages such as increased spectral efficiency, better energy efficiency, and improved interference management.
6. *Enhanced Diversity and Robustness:* Beyond 4×4 MIMO, additional antennas can provide enhanced diversity, improving the system's robustness against fading and interference. This is especially valuable in dynamic and challenging radio environments.
7. *5G Evolution:* The evolution of 5G networks involves the use of advanced MIMO configurations, including 8×8 MIMO and even higher. These configurations contribute to achieving the ambitious goals of 5G, such as ultra-reliable low-latency communication (URLLC) and massive machine-type communication (mMTC).
8. *Customization for Specific Use Cases:* In certain applications, such as specialized industrial applications or mission-critical communications, deploying more antennas allows for customization of the MIMO system to meet specific performance requirements.

While the benefits of going beyond 4×4 MIMO are substantial; practical considerations, including cost, complexity, and spectrum availability, must be taken into account. The choice of MIMO configuration depends on the specific use case, deployment scenario, and the trade-offs between performance gains and implementation challenges.

7.5 SIGNAL PROCESSING CHALLENGES IN MIMO SYSTEMS

The main signal processing challenges in MIMO systems result from the massive antenna configuration and the large path loss, which make traditional linear channel estimation methods ineffective and the popular assumption in MIMO-NOMA, e.g., perfect instantaneous channel state information (CSI) at the base station (BS), impractical. Non-Orthogonal Multiple Access (NOMA), is the latest development in the multiple access technology; and it is envisioned to be an essential component of 5G mobile networks. The combination of NOMA and multi-antenna multi input multi output (MIMO) technologies exhibits a significant potential in improving spectral efficiency and providing better wireless services to more users.

For Orthogonal Multiple Access (OMA), a main issue is the low spectral efficiency when the wireless (time/frequency/code) resources are allocated to users with poor channel conditions. NOMA alleviates this problem by using Superposition Coding (SC) and Successive Interference Cancellation (SIC) and enabling users with significantly different channel conditions to share the same resource block. Thus, NOMA is regarded as one of the enabling technologies to meet the requirements of the high rate, dense coverage, massive connectivity, and low latency in 5G wireless systems. In NOMA, a good balance between the spectral efficiency and the user fairness can be achieved via proper user pairing and power allocation policies, with affordable costs in SIC complexity and signaling overhead.

7.6   MATLAB COMMUNICATIONS TOOLBOX

Several handy built-in functions and blocks available in the Communications Tool-box/Simulink, for simulating the performance of MIMO systems; including massive MIMO. We will consider some of them in this section. A more detailed information can be obtained from MATLAB online/offline references.

1. MIMO-OSTBC, sphere decoding, and massive MIMO–OSTBC Encoder and OSTBC Combiner blocks in Simulink; the comm.MIMOChannel system object or the MIMO Fading Channel block in Simulink. We can model a sphere decoder using the comm.SphereDecoder System object in MATLAB or the Sphere Decoder block in Simulink.
2. MIMO Fading Channel Block. The MIMO Fading Channel block filters an input signal using a MIMO multipath fading channel. One can model a MIMO fading channel using the comm.MIMOChannel System object in MATLAB or the MIMO Fading Channel block in Simulink.
3. MIMO Transfer functions.
4. Tools for analysis of MIMO models.
5. MIMO State Space Models–show how to represent MIMO systems as state-space models.
6. Massive MIMO Hybrid Beamforming; shows how hybrid beamforming is employed at the transmit end of a massive MIMO communications system, using techniques for both multi-user and single-user systems.
7. OFDM with MIMO Simulation– shows how to use an OFDM modulator and demodulator in a simple, 2×2 MIMO error rate simulation.

7.7   CONCLUDING REMARKS

The development of MIMO communication systems has revolutionized wireless communication, providing increased data rates and reliability. Simulating MIMO systems using MATLAB and Python allows engineers and researchers to experiment with different configurations and channel conditions, facilitating the design and op-timization of MIMO-enabled communication systems for various applications. The code snippets given in this chapter offer a starting point for implementing MIMO simulations, serving as a practical guide for those interested in exploring the capabilities of MIMO technology.

Application of MIMO techniques for sending and receiving multiple data signals simultaneously over the same radio channel by exploiting multipath propagation that provide potential gains in capacity when using multiple antennas at both transmitter and receiver ends of a communications system. New techniques, which account for the extra spatial dimension, have been adopted to realize these gains in new systems and previously existing systems. MIMO technology has been adopted in multiple wireless systems, including WiFi, Worldwide Interoperability for Microwave Access (WiMAX), LTE, and LTE-Advanced.

FURTHER READING

1. Vahid Tarokh, Hamid Jafarkhani, and A. R. Calderbank, Space-Time Block Codes from Orthogonal Designs, *IEEE Transactions on Information Theory*, Vol. 45, No. 5, July 1999.
2. Robert W. Heath, et al. An Overview of Signal Processing Techniques for Millimeter Wave MIMO Systems, *IEEE Journal of Selected Topics in Signal Processing*, Vol. 10, No. 3, April 2016.
3. D. Tse and P. Viswanath, *Fundamentals of Wireless Communications,* Cambridge University Press, 2005.
4. S.M. Alamouti, A simple transmit diversity technique for wireless communications, *IEEE Journal on Selected Areas in Communications,* Vol. 16, No. 8, October 1998.
5. Q.H. Spencer, et al., Zero-Forcing Methods for Downlink Spatial Multiplexing in Multiuser MIMO Channels, *IEEE Transactions on Signal Processing,* Vol. 52, No. 2, February 2004.
6. Yongming Huang, et al., Signal Processing for MIMO-NOMA: Present and Future Challenges, *IEEE Wireless Communications*, Vol. 25, No. 2, April 2018.
7. The MathWorks Inc., *Communications Toolbox Reference*, 2023.
8. The MathWorks Inc., *Signal Processing Toolbox User Guide*, 2023

EXERCISES

1. How does the performance of a MIMO system vary with different modulation schemes such as QPSK, 16-QAM, and 64-QAM?
2. What is the impact of varying the number of transmit and receive antennas on the Bit Error Rate (BER) performance of a MIMO system?
3. How does changing the spacing between antennas affect the capacity of a MIMO system?
4. Investigate the diversity gain of a MIMO system under different fading conditions (e.g., Rayleigh, Rician).
5. Explore the impact of spatial correlation between antennas on the performance of a MIMO system.
6. How does beamforming or precoding techniques enhance the performance of a MIMO system in terms of signal-to-noise ratio (SNR) and spectral efficiency?
7. Compare the performance of different MIMO detection techniques (e.g., Zero Forcing (ZF), Minimum Mean Square Error (MMSE), and Maximum Likelihood Estimation (MLE)) in terms of complexity and BER performance.
8. Investigate the impact of channel estimation errors on the performance of a MIMO system and evaluate robust techniques to mitigate these errors.
9. How does the use of spatial multiplexing techniques such as Vertical Bell Labs Layered Space-Time (V-BLAST) or Orthogonal Space-Time Block Coding (OSTBC) improve the data rate of a MIMO system?
10. Explore the trade-offs between diversity gain and spatial multiplexing gain in MIMO systems under different channel conditions.

8 Signal Processing in Wireless Communication

We will discuss the signal-processing aspects of various wireless communication systems in this chapter. Wireless communication technology has matured over the past several decades. The credit for developing the first wireless communication system goes to Italian physicist and inventor, Guglielmo Marconi, who developed the first practical system of wireless telegraphy, in 1895. Later several inventors worked on developing a practical telephone system, and were successful in developing practical wireless telephone systems; even though the first practical telephone system developed by American engineer Alexander Graham Bell was wired.

We will mainly consider the signal processing aspects of wireless communication systems, including the mobile phones; their several generations. The credit for developing the first successful mobile (wireless) phone goes to Martin Cooper, who was an engineer at Motorola. Cooper made the first hand held cellular mobile phone call on April 3, 1973. He used a prototype of the Motorola DynaTAC (Dynamic Adaptive Total Area Coverage) phone, which weighed about 1.1 kg and had a battery life of about 30 minutes. Cooper made the historic call from a sidewalk in New York City to his rival at Bell Labs, Joel Engel, who was also working on developing cellular technology. This demonstration marked a significant milestone in the history of telecommunications, paving the way for the widespread adoption of mobile phones and the subsequent evolution of mobile communication technologies.

8.1 INTRODUCTION

Mobile cellular communication systems have revolutionized the way people communicate and interact with one another, providing ubiquitous connectivity and enabling seamless communication on the go. These systems form the backbone of modern telecommunications networks, facilitating voice calls, data messaging, internet access, and various other services like Multi Media Messaging (MMS), through wireless connections.

The core of the mobile cellular communication systems is the concept of cellular networks, which divide geographical areas into smaller regions called cells. Each cell is served by a base station, also known as a cell site or cell tower, which communicates with mobile devices within its coverage area. The use of cells allows for efficient frequency reuse and increases the capacity and coverage of the network.

DOI: 10.1201/9781003527589-8

8.2 MOBILE CELLULAR WIRELESS COMMUNICATION

The evolution of mobile cellular communication systems can be traced back to the early analog systems introduced in the 1980s, such as the Advanced Mobile Phone System (AMPS) in the United States and the Total Access Communication System (TACS) in Europe. These systems provided basic voice communication services and operated on relatively low frequencies.

The transition to digital cellular systems in the 1990s marked a significant advancement in mobile communications, enabling improved voice quality, increased capacity, and support for new services. The Global System for Mobile Communications (GSM) standard, developed in Europe, became widely adopted and standardized digital cellular communication system worldwide. GSM introduced features such as encryption, international roaming, and support for text messaging (SMS). As mobile data usage grew rapidly in the early 2000s, the introduction of 3rd Generation (3G) and subsequent 4th-Generation (4G) cellular technologies brought higher data speeds and enhanced multimedia capabilities to mobile devices. Technologies such as Universal Mobile Telecommunications System (UMTS) and Long-Term Evolution (LTE) enabled faster internet access, video streaming, and other data-intensive applications.

Today, mobile cellular communication systems continue to evolve with the deployment of 5th Generation (5G) and 6G technology. 5G/6G promise even higher data rates, ultra-low latency, massive device connectivity, and support for emerging applications such as Internet of Things (IoT), augmented reality (AR), and virtual reality (VR). With the deployment of 5G and 6G networks, mobile communication is expected to become more versatile, efficient, and pervasive, driving innovation across various industries.

8.2.1 SUBSYSTEMS OF MOBILE CELLULAR COMMUNICATION SYSTEM

The key subsystems of mobile cellular communication systems are:

1. *Mobile Devices:* Including Smartphones, tablets, and other mobile devices equipped with wireless communication capabilities form the end-user interface of the cellular network. These devices support voice calls, messaging, internet browsing, and access to various applications and services.
2. *Base Transceiver Stations:* Base stations (BTS), also known as cell sites or cell towers, are deployed throughout the coverage area to provide wireless connectivity to mobile devices. They communicate with mobile devices within their respective cells and facilitate handovers as users move between cells.
3. *Core Network:* The core network manages the routing of voice and data traffic between base stations, mobile devices, and external networks such as the Public Switched Telephone Network (PSTN) and the internet. It includes components such as mobile switching centers (MSCs), home location registers (HLRs), and serving gateways (SGWs).
4. *Radio Access Network (RAN):* The RAN encompasses the radio access technologies and protocols used to establish wireless connections between mobile devices

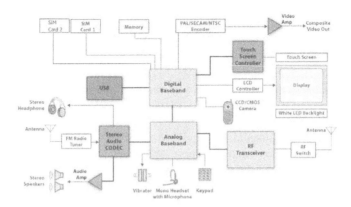

Figure 8.1 Block Diagram of a Mobile Phone.

and base stations. It includes components such as radio transceivers, antennas, and radio frequency (RF) equipment.

5. *Backhaul Network:* The backhaul network provides connectivity between base stations and the core network, typically using wired or wireless links. It carries traffic between base stations and central network elements such as MSCs and data centers.

The block diagram of a typical mobile phone is shown in figure 8.1.

8.3 INTRODUCTION TO DIGITAL CELLULAR MOBILE COMMUNICATION

The modern digital cellular mobile systems, including the 5G and 6G systems, are improved versions of the first, immensely popular digital mobile phone system, the Global System for Mobile (GSM). We will now study the various generations of mobile phone systems and GSM in greater detail.

8.3.1 GENERATIONS OF CELLULAR MOBILE TECHNOLOGY

The first generation (1G) mobile systems includes the following standards, which are basically analog cellular portable radiotelephone standards. They are: Nordic Mobile Telephone (NMT); Advanced Mobile Phone System (AMPS); Total Access Communication System (TACS), which is the European version of AMPS; and the Japanese Total Access Communication System (JTACS) developed by NTT DoCoMo.

The 2G standards include Global System for Mobile Communications (GSM); Integrated Digital Enhanced Network (iDEN); Digital Advanced Mobile Phone System (D-AMPS) based on TDMA; Code Division Multiple Access (CDMA One) defined by IS-95 standard; and Personal Digital Cellular (PDC). The 2.5G mobiles include a set of transition technologies between 2G and 3G wireless technologies. In addition to voice, it involves digital communication technologies that support E-mail and simple Web browsing. The following are considered to be part of 2.5G mobile systems: General Packet Radio Service (GPRS) and Wideband Integrated Dispatch Enhanced Network (WiDEN). The 2.75G includes CDMA 2000 and Enhanced Data rates for GSM Evolution (EDGE).

The 3G mobiles supports broadband voice, data, and multimedia communication technologies in wireless networks. It includes Wideband Code Division Multiple Access (W-CDMA); Universal Mobile Telecommunications System (UMTS); Freedom of Mobile Multimedia Access (FOMA); CDMA2000 1xEV which is more advanced than CDMA2000, and it supports 1xEV technology and can meet 3G requirements and Time Division-Synchronous Code Division Multiple Access (TD-SCDMA). 3.5G refers to a technology that goes beyond the development of comprehensive 3G wireless and mobile technologies and include High-Speed Downlink Packet Access (HSDPA). The 3.75G technology goes beyond the development of comprehensive 3G wireless and mobile technologies and include High-Speed Uplink Packet Access (HSUPA).

The 4G mobiles implement high-speed mobile wireless communications technology and designed to enable new data services and interactive TV services in mobile networks. 4G is the current mainstream cellular service offered to cell phone users, performance roughly 10× faster than 3G service. One of the most important features in the 4G mobile networks is the domination of high-speed packet transmissions or burst traffic in the channels

5G mobile systems offer lower response times (lower latency) and higher data transfer speeds.The performance goals of 5G systems are high data rates, reduced latency, energy savings, reduced costs, increased system capacity, and large-scale device connectivity. 5G is still a fairly new type of networking and is still being spread across nations. Moving forward, 5G is going to set the standard of cellular service around the whole globe.

Now, the 6G mobile systems are in development stage and are supposed to be a successor to the 5G technology. 6G is being developed by numerous companies including Airtel, Anritsu, Apple, Ericsson, Fly, Huawei, Jio, Keysight, LG, Nokia, NTT DoCoMo, Samsung, Vi, and Xiaomi. 6G networks will likely be significantly faster than previous generations, and are expected to be more diverse; and are likely to support applications beyond current mobile use scenarios, such as ubiquitous instant communications, pervasive intelligence, and the Internet of Things (IoT).

8.3.2   GLOBAL SYSTEM FOR MOBILE (GSM)

The first generation (1G) mobile systems introduced analog voice calls in the 1980s, paving the way for basic wireless communication. The second generation (2G) mobile systems brought in digital networks in the 1990s, enabling SMS messaging and basic data services. The GSM is considered to be a 2G Mobile system. With the advent of the third generation (3G) in the early 2000s, mobile internet access became feasible, allowing for faster data transfer speeds and multimedia capabilities.

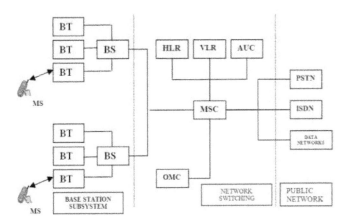

Figure 8.2 Block Diagram of a GSM Network.

The fourth generation (4G) further enhanced data speeds and introduced technologies like Long Time Evolution (LTE), enabling high-definition video streaming and advanced mobile applications. Currently, the fifth generation (5G) is revolutionizing connectivity with ultra-fast speeds, low latency, and massive device connectivity, unlocking the potential for innovations like autonomous vehicles, augmented reality, and the Internet of Things (IoT).

Digital cellular systems incorporate digital modulation techniques for their functioning. Digital systems provide significant improvements in capacity and system performance. The United States Digital Cellular System (USDC) was created in the late 1980s to handle more users in a given spectrum allotment. USDC is a time division multiple access (TDMA) technology. USDC can provide up to six times the capacity of Advanced Mobile Phone System (AMPS); which was developed by Bell Labs from 1968 to 1983; and became the first cellular network standard in the United States. It was an analog mobile phone system.

The block diagram of a GSM network is shown in figure 8.2.

8.3.3   GENERAL PACKET RADIO SERVICE (GPRS)

The General packet radio service (GPRS) is a mobile communications standard that operates on 2G and 3G cellular networks to enable moderately high-speed data transfers using packet-based technologies. GPRS was developed by European Telecommunications Standards Institute (ETSI) because of the prior Cellular Digital Packet Data (CDPD) service, and the advancements in I-mode packet switched cell services. GPRS overrides the wired networks, as this framework has streamlined access to the packet information network like the internet. The packet radio standard is utilized by GPRS to transport client information packets in a structured route between GSM versatile stations and external packet information networks. These packets can be straightforwardly directed to the packet-changed systems from the GPRS portable stations.

The services included in GPRS are SMS messaging and broadcasting, push-to-talk over cellular, Instant messaging and presence, Multimedia messaging service (MMS), and Point-to-Point and Point-to-Multipoint services. GPRS support the Internet Protocol (IP) and the Point-To-Point Protocol (PPP).

8.4   MITIGATION OF CCI AND ACI

Mobile cellular channels are often modeled as Linear Time-Variant (LTV) channels. In the case of mobile cellular systems, the co-channel interference (CCI) and adjacent channel interference (ACI) are two factors that affect the performance of the system.Let us study CCI and ACI detail in the following sections.

8.4.1   CO-CHANNEL INTERFERENCE

The co-channel interference results from the reuse of frequencies for the forward channel (uplink) and reverse channel (downlink). In the cellular mobile system, there is actually a need for frequency reuse in channels due to the limited spectrum availability.

8.4.1.1   Frequency Reuse/Frequency Planning

Mobile cellular systems rely on an intelligent allocation and reuse of frequencies allotted for channels throughout a coverage area. Each cellular base station (BTS) is allocated a group of radio channels to be used within a small geographic area called a cell. Consider a a cellular system which has a total of S duplex channels available for use. If each cell is allocated a group of k channels such that $k < S$, and is the S channels are divided among N cells into unique and disjoint channel groups which, each has the same number of channels, the total number of available radio channels is given by $S = kN$. The N cells which collectively use the complete set of available frequencies are called a *cluster*. If a cluster is replicated M times within the system, the total number of duplex channels, C, can be used as a measure of capacity and is given by $C = MkN = MS$. Base stations in adjacent cells are assigned channel groups which contain completely different channels than the neighboring cells. The base station antennas are designed to achieve the desired coverage within the particular cell. By limiting coverage to within the boundaries of a cell, the same group of channels (frequencies) can be used to cover different cells that are separated from one another by distances large enough to keep the interference levels within tolerable limits. This concept is called *frequency reuse or frequency planning*. Frequency reuse implies that in a given coverage area there are several cells that use the same set of frequencies.

By using the hexagon geometry for the cells, the fewest number of them can cover a geographic region. Moreover, the hexagon geometry closely approximates the circular radiation pattern for a *omni-directional* base station antenna. It can be shown that when the cluster size, N is 7, the frequency reuse factor is $1/7$. In order to connect without gaps between adjacent cells, the geometry of the hexagon is such that the number of cells per cluster, N, can only take values which satisfy equation 8.4.1, where i and j are nonnegative numbers.

$$N = i^2 + ij + j^2 \qquad (8.4.1)$$

To find the nearest co-channel neighbors of a particular cell, one must move i cells along any chain of hexagons and then turn 60 degrees counter-clockwise and then move j cells.

8.4.1.2 Co-Channel Interference and System Capacity

With frequency reuse, there are several cells that use the same frequency, which are known as *co-channels*. The interference between signals from these cells is termed as *Co-Channel Interference*, To reduce CCI, co-channel cells must be physically separated by a minimum distance to provide sufficient isolation. When the size of each cell is nearly the same, and the base stations transmit the same power, the CCI ratio is independent of the transmitted power and becomes a function of the radius of the cell (R) and the distance between centers of the nearest co-channel cells (D). By increasing the D/R ratio, the spatial separation between co-channel cells in relation to the coverage distance of a cell is increased. Thus CCI is reduced due to the improved isolation of RF energy from the co-channel cell. The parameter Q, called the *co-channel reuse ratio*, is related to the cluster size. For the hexagonal geometry, $Q = D/R = \sqrt{3N}$. A small value of Q provides larger capacity since the cluster size N is small, where as a larger value of Q improves the transmission quality, due to a smaller value of CCI. Hence, a trade-off between these two must be made in the actual design.

8.4.1.3 Signal-to-Interference Ratio (SIR)

If i_0 is the number of interfering channels, then the SIR for a mobile receiver can be expressed as

$$SIR = \frac{S}{\sum\limits_{i=0}^{i_0} I_i} \qquad (8.4.2)$$

Measurements on propagation show that in a mobile radio channel, the average received power P_r at a distance d from the transmitting antenna is approximately:

$$P_r = P_0 \left(\frac{d}{d_0} \right)^{-n} \qquad (8.4.3)$$

where P_0 is the power received at a close-in reference point in the far field region of the antenna, at a distance d_0 from the transmitting antenna, and n is the path loss exponent. The value of n varies from 2 to 4 in urban cellular systems. If D_i is the distance of the ith interferer from the mobile, the received power at a given mobile due to the ith interfering cell will be proportional to D_i^{-n}. When the transmit power of each base station is equal and the path loss exponent is the same throughout the coverage area, the SIR for a mobile can be approximated as:

$$SIR = \frac{R^{-n}}{\sum_{i=0}^{i_0} D_i^{-n}} \qquad (8.4.4)$$

If we consider only the first layer of interfering cells, and if all the interfering base stations are equidistant from the desired base station, and also if this distance is equal to the distance D between cell centers, then equation 8.4.4 simplifies to

$$SIR = \frac{(D/R)^n}{i_0} = \frac{(\sqrt{3N})^n}{i_0} = \frac{Q^n}{i_0} \qquad (8.4.5)$$

Using an exact cell geometry layout, it can be shown for a 7-cell cluster (N=7), with mobile unit at the cell boundary, the mobile is at a distance $D-R$ from the two nearest co-channel interfering cells and is exactly $D+R/2$, D, $D-R/2$, and $D+R$ from the other interfering cells in the first tier. Hence equation 8.4.4 can be approximated as (assuming the path loss exponent, $n=4$)

$$SIR = \frac{1}{2 \times [(Q-1)^{-4} + (Q+1)^{-4} + Q^{-4}]} \qquad (8.4.6)$$

For $N=7$, the co-channel reuse ratio, $Q = \frac{D}{R} = \sqrt{3N} = 4.6$, and the worst-case SIR is exactly 49.56 (16.95dB). A plot of N versus SIR is shown in figure 8.3. The MATLAB/GNU Octave code to generate the same is given below.

```
%% MATLAB/GNU Octave Script to
% plot Cluster size (N) versus SIR.
close; close;
N=[];
for i=1:25
    for j=1:25
        N=[N,i^2+i*j+j^2];
    end;
end;
N=sort(N);
Q=sqrt(3*N);
sir=1./(2*((Q-1).^(-4)+Q.^(-4)+(Q+1).^(-4)));
sirdb=10*log10(sir);
plot(N,sirdb,LineWidth=2);
xlabel('Cluster Size, N');
ylabel('SIR in dBs');
title('Cluster Size Versus SIR in dBs');
grid;
```

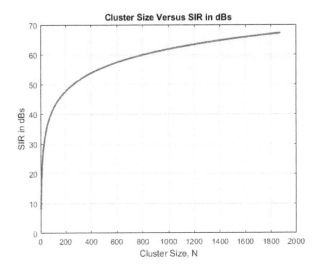

Figure 8.3   Plot of Cluster Size Versus SIR in dBs.

Subjective tests indicate that for sufficient voice quality, a $SIR \geq 18dB$ is needed. Hence to ensure sufficient voice quality, it is necessary to increase N to the next larger size, which from equation 8.4.2 is found to be 12; corresponding to $i = j = 2$. This will obviously result in a significant decrease in capacity, since 12-cell reuse offers a spectrum utilization of 1/12 within each cell. From the above discussion, *it is evident that CCI determines link performance, which in turn dictates the frequency reuse plan and the overall capacity of cellular systems.*

8.4.2   ADJACENT CHANNEL INTERFERENCE

The interference from signals which are adjacent in frequency to the desired signal is called *Adjacent Channel Interference (ACI)*. ACI results from imperfect filters at the receiver, which allows nearby frequencies leak into the passband. The ACI can be serious if an adjacent channel user is transmitting in very close range to a subscriber's receiver, while the receiver is attempting to receive a base station on the desired channel. This is called the *near-far effect*, where a nearby transmitter (which may or may not be of the same type as that used by the cellular system) captures the receiver of the subscriber. Also when a mobile close to a base station transmits on a channel close to one being used by a weak mobile, the near-far effect occurs.

The ACI can be mitigated through careful filtering and channel assignments. Since each cell is given only a fraction of the available channels, a cell need not be assigned channels which are adjacent in frequency. By keeping the separation of frequency between each channel in a given cell as large as possible, which form a contiguous band of frequencies within a particular cell, channels are allocated in such a way that the separation in frequency between adjacent channels is maximized.

There exist many channel allocation schemes that are able to separate adjacent channels in a cell by as many as N channel bandwidths, where N is the cluster size.

If the frequency reuse factor is large (e.g., small N), the separation between adjacent channels at the base station may not be sufficient to keep the ACI level within tolerable limits. As an example, a close-in mobile is 20 times as close as to the base station as another mobile and has energy S, the spill out of its passband, the SIR at the base station for the weak mobile (before filtering at the receiver) is approximately:

$$\frac{S}{I} = 20^{-n} \tag{8.4.7}$$

where n is the path loss exponent. Note that for $n = 4$, SIR is $-52dB$.

8.5   INTER-SYMBOL INTERFERENCE (ISI)

One of the biggest problems in the field of mobile cellular communication channels is inter-symbol interference (ISI). With the increasing amount of time we spend on mobile devices, it is critical to comprehend the complexities of ISI in order to maximize network performance and guarantee smooth user experiences when communicating.

Modern communication systems are based on mobile cellular channels, which allow voice, data, and multimedia content to be transmitted between mobile devices and network infrastructure. Numerous impairments, like as noise, fading, and interference, can affect the dependability of communication and deteriorate the quality of the signal transmitted across these channels.

ISI is the result of symbols representing bits of information overlapping with one another while being communicated over a communication channel. This distortion makes it difficult for the recipient to decode the data correctly. In mobile wireless channels, with multipath propagation, when signals follow several pathways and arrive at the receiver with varying delays, this effect is most noticeable.

Multipath propagation is a prevalent phenomenon in mobile cellular networks because of environmental impediments such as buildings and topography that produce signal reflections, diffraction, and scattering. Because of this, signals sent from the base station to the mobile device may travel through several channels before reaching the receiver, each with a unique delay and degree of attenuation.

Symbols transferred in one-time interval may interfere with symbols transmitted in subsequent intervals due to the variable delays of the multipath components, resulting in ISI. The received signal may be distorted as a result of this interference, making it more challenging for the receiver to correctly decipher the transmitted data.

Several factors can contribute to the severity of ISI in mobile cellular channels:

1. *Symbol Duration:* In digital communication systems, symbols are typically transmitted over a finite duration of time. The duration of a symbol depends on the modulation scheme and the signaling rate of the system. Shorter symbol durations increase the susceptibility to ISI since there is less temporal separation between adjacent symbols.

2. *Channel Characteristics:* The characteristics of the mobile cellular channel, such as delay spread and Doppler spread, play a significant role in determining the severity of ISI. Delay spread refers to the difference in arrival times between the earliest and latest multipath components, while Doppler spread accounts for the frequency shift caused by the motion of the mobile device or the base station. Channels with larger delay spread and Doppler spread are more prone to ISI.

3. *Modulation Scheme:* The choice of modulation scheme affects the resilience of the communication system to ISI. For instance, higher-order modulation schemes, such as Quadrature Amplitude Modulation (QAM), allow for higher data rates but are more susceptible to ISI compared to lower-order modulation schemes like Binary Phase Shift Keying (BPSK) or Quadrature Phase Shift Keying (QPSK).

8.5.1 MITIGATION OF ISI

Efficient mitigation techniques are essential to combat ISI and improve the reliability of mobile cellular communication channels. Some common methods are:

1. *Equalization:* Equalization techniques are used at the receiver to compensate for the effects of ISI by applying appropriate filtering to the received signal. Linear equalizers, such as zero-forcing equalizers and minimum mean square error (MMSE) equalizers, attempt to invert the channel response to recover the transmitted symbols accurately. However, linear equalizers may amplify noise and lead to error propagation in certain cases.

2. *Adaptive Equalization:* Adaptive equalization algorithms dynamically adjust the equalizer coefficients based on the characteristics of the channel and the received signal. Adaptive algorithms, such as the Least Mean Squares (LMS) algorithm and the Recursive Least Squares (RLS) algorithm, continuously update the equalizer parameters to track changes in the channel conditions and minimize ISI.

3. *Diversity Techniques:* Diversity techniques exploit the spatial, temporal, or frequency diversity inherent in the communication channel to combat the effects of fading and ISI. Spatial diversity involves using multiple antennas at the transmitter or receiver to capture independent fading paths, while temporal diversity exploits variations in the channel over time. Frequency diversity, on the other hand, utilizes multiple frequency channels to transmit redundant copies of the data, which can be combined at the receiver to mitigate ISI and fading.

4. *Orthogonal Frequency Division Multiplexing (OFDM):* OFDM is a modulation technique that divides the available bandwidth into multiple subcarriers, each carrying a portion of the data. By spacing the subcarriers closely together, OFDM reduces the symbol duration and mitigates the effects of ISI caused by multipath propagation. Additionally, the use of cyclic prefix (CP) in OFDM provides a guard interval between adjacent symbols, further reducing ISI. Additionally, the use of a cyclic prefix (CP) in OFDM provides a guard interval between adjacent symbols, further reducing ISI.

Precisely, Inter Symbol Interference (ISI) poses a significant challenge in mobile cellular communication channels, particularly in environments with multipath propagation. Understanding the factors contributing to ISI and employing efficient mitigation techniques are essential for optimizing network performance and ensuring reliable communication experiences for mobile users. As mobile wireless technology continues to evolve, ongoing research and innovation in ISI mitigation will play a crucial role in advancing the capabilities of mobile cellular networks and meeting the growing demands of modern communication systems.

8.6   CHANNEL EQUALIZATION

In mobile wireless communication, channel equalization techniques play a major role in ensuring robust and reliable data transmission. As signals travel from transmitters to receivers, in a wireless manner, they encounter various obstacles such as multipath fading, interference, and noise. These factors distort the transmitted signal, leading to errors in data reception. Channel equalization techniques are employed to mitigate these effects and recover the original transmitted signal, thereby enhancing the performance of mobile wireless systems.

Multipath fading, a common phenomenon in wireless communication, occurs when signals take multiple paths to reach the receiver due to reflections, diffractions, and scattering from surrounding objects. As a result, the transmitted signal arrives at the receiver with different delays and phases, causing intersymbol interference (ISI). ISI occurs when symbols transmitted in one-time interval overlap with symbols from adjacent intervals, making it challenging for the receiver to correctly decode the received data.

One of the primary channel equalization techniques used to combat ISI is *linear equalization*. Linear equalizers, such as zero-forcing equalizers and minimum mean square error (MMSE) equalizers, apply filters to the received signal to compensate for the channel distortion. Zero-forcing equalizers attempt to invert the channel response to eliminate ISI entirely, while MMSE equalizers minimize the mean square error between the transmitted and received symbols, taking into account the channel characteristics and noise.

Adaptive equalization algorithms take channel variations into account and adjust the equalizer parameters dynamically to track changes in the channel conditions. Algorithms such as the Least Mean Squares (LMS) and Recursive Least Squares (RLS) algorithms continuously update the equalizer coefficients based on the received signal and feedback from the receiver. Adaptive equalization techniques are particularly effective in mobile wireless communication systems, where channel conditions are subject to rapid changes due to user mobility and environmental factors.

To summarize, channel equalization techniques play a critical role in enhancing the performance of mobile wireless communication systems by mitigating the effects of multipath fading, interference, and noise. Linear equalization, adaptive equalization, diversity techniques, and OFDM are among the key approaches used to combat intersymbol interference and improve the reliability and efficiency of wireless communication channels. As mobile communication technologies continue to evolve, ongoing research and development efforts are focused on further optimizing these techniques to meet the growing demands of mobile users for faster, more reliable, and ubiquitous connectivity.

8.7   NON-ORTHOGONAL MULTIPLE ACCESS (NOMA)

Non-Orthogonal Multiple Access is a technology used in mobile wireless communication systems to improve spectral efficiency and enhance the overall performance of the network. In traditional wireless communication systems, multiple users access the same frequency resources using orthogonal multiple access (OMA) techniques such as Frequency Division Multiple Access (FDMA), Time Division Multiple Access (TDMA), or Code Division Multiple Access (CDMA). These techniques allocate separate frequency bands, time slots, or codes respectively, to different users to avoid interference between transmissions.

In contrast, NOMA allows multiple users to share the same frequency, time, and code resources simultaneously by exploiting the power domain for multiplexing. Instead of allocating orthogonal resources to different users, NOMA assigns different power levels to each user, allowing them to share the same resource block. Users with better channel conditions are allocated higher power levels, while users with poorer channel conditions are assigned lower power levels. The key points behind NOMA include:

- *Power Domain Multiplexing:* NOMA multiplexes multiple users in the power domain by allocating different power levels to each user. Users with stronger channel conditions are assigned higher power levels, while users with weaker channel conditions receive lower power levels.
- *Successive Interference Cancellation (SIC):* At the receiver, NOMA utilizes successive interference cancellation techniques to decode the signals from multiple users. The receiver decodes the signal of the user with the highest power level first and subtracts it from the received signal before decoding the signal of the next user with a lower power level. This process continues iteratively until all users' signals are decoded.

NOMA can offer several merits over traditional OMA techniques:

- *Spectral Efficiency:* By allowing multiple users to share the same frequency resources simultaneously, NOMA improves spectral efficiency and increases the capacity of the communication system.
- *Fairness:* NOMA provides a fairer resource allocation scheme compared to OMA techniques, as it allocates resources based on the channel conditions of users. Users with better channel conditions are allocated higher power levels, ensuring that all users receive adequate quality of service.
- *Flexibility:* NOMA is a flexible and adaptive technology that can be easily integrated into existing wireless communication systems without requiring significant changes to the infrastructure.
- *Enhanced Coverage and Throughput:* NOMA can improve coverage and throughput in cellular networks, especially in dense urban environments

and areas with high user densities, by efficiently utilizing the available spectrum and mitigating interference.

NOMA has gained significant attention in recent years as a promising technology for future wireless communication systems, including 5G and beyond. Its ability to improve spectral efficiency, enhance system capacity, and provide fair resource allocation makes it a compelling solution for meeting the increasing demands of mobile wireless networks. Ongoing research and development efforts are focused on further optimizing NOMA techniques and integrating them into commercial communication systems to realize the full potential of this innovative technology.

We will now consider some of the challenges/issues faced by the NOMA technology.

1. *The Near-Far Problem:* NOMA relies on power-domain multiplexing, where users with different channel conditions share the same frequency band. This can lead to the near-far problem, where users close to the base station (BS) may dominate the channel resources, causing degradation in the performance of users farther away.
2. *System Complexity and Cost of Implementation:* Implementing NOMA requires sophisticated signal processing techniques both at the transmitter and receiver ends, which can increase system complexity and cost.
3. *Interference:* NOMA can introduce intra-cell interference among users sharing the same resource block. Managing this interference becomes crucial, especially in dense deployment situations, to maintain the desired quality of service.
4. *Scheduling Challenges:* Efficient user scheduling is crucial in NOMA systems to maximize spectral efficiency. However, scheduling users with varying channel conditions and quality of service requirements in NOMA can be challenging, requiring complex algorithms.
5. *Fairness Issues:* NOMA inherently prioritizes users with better channel conditions, potentially leading to fairness issues, where users with poor channel conditions may experience lower quality of service or throughput degradation.
6. *Backward Compatibility:* Introducing NOMA into existing communication systems may require significant upgrades or even replacement of entire infrastructure, which can be costly and may face compatibility issues with legacy devices.
7. *Power Allocation:* Optimal power allocation among users is critical in NOMA to achieve maximum throughput. However, determining the appropriate power allocation strategy can be complex, especially in dynamic and heterogeneous networks and disparate channel conditions.
8. *Security Challenges:* NOMA introduces new security challenges, particularly concerning user privacy and confidentiality. Since multiple users share the same frequency band, ensuring secure communication and preventing eavesdropping becomes more challenging.
9. *Performance Variability:* NOMA performance heavily depends on the channel conditions and the distribution of users in the cell. Variability in these factors can lead to unpredictable performance fluctuations.

8.8   CONCLUDING REMARKS

We discussed the signal-processing aspects of various wireless communication systems in this chapter. Wireless communication technology has matured over the past several decades. Signal processing plays a pivotal role in advancing wireless communication systems. This chapter has delved into the key techniques and challenges within this domain, showcasing its critical importance in enabling efficient data transmission, reception, and interpretation. We discussed the modulation schemes, channel estimation, and equalization, that contribute to enhancing spectral efficiency, improving error rates, and maximizing throughput in wireless networks. As technology evolves, signal processing continues to evolve alongside, addressing emerging challenges such as interference mitigation, resource allocation, and the integration of advanced technologies like NOMA. By fostering a deeper understanding of signal processing principles and methodologies, this chapter empowers researchers and practitioners to innovate and drive the next wave of advancements in wireless communication, shaping the future of connectivity and enabling new possibilities in the realm of wireless applications.

FURTHER READING

1. K. C. Raveendranathan, *Neuro-Fuzzy Equalizers for Mobile Cellular Channels,* CRC Press, 2014.
2. Theodore S. Rappaport, *Wireless Communication: Principles and Practice,* 2nd Edition, Pearson Education, 2022.
3. Yongming Huang et al., Signal Processing for MIMO-NOMA: Present and Future Challenges, *IEEE Wireless Communications,* Vol. 25, No. 2, April 2018.
4. Vijay K. Garg, *Wireless Communications and Networking,* Morgan Kaufmann Publishers, 2007.

EXERCISES

1. How does signal processing contribute to improving the spectral efficiency of wireless communication systems?
2. What are the main challenges in channel estimation and equalization in wireless communication, and how are they addressed through signal processing techniques?
3. Explain the role of modulation schemes in wireless communication, and how signal processing optimizes their performance.
4. List some advanced signal processing methods used for interference mitigation in wireless communication networks.
5. Explain how signal processing enables efficient resource allocation and scheduling in wireless communication systems.
6. How the near-far problem in NOMA-based wireless communication is addressed using signal processing.

7. How do adaptive signal processing algorithms improve the performance and adaptability of wireless communication systems in dynamic environments?
8. Discuss the impact of signal processing on reducing latency and enhancing reliability in wireless communication for real-time applications.
9. What are the recent advancements in signal processing techniques for enhancing security and privacy in wireless communication networks?
10. How does signal processing contribute to the development of emerging technologies such as massive MIMO and mmWave communication in wireless networks?

9 Advanced Communication Systems

Recently there has been considerable, extensive developments taking place in the domain of advanced communication systems. These include several novel multiplexing techniques, modulation schemes, and noise mitigation schemes. All these developments resulted in better reception of signals, connectivity, better throughput, and low latency. The subject matter of this chapter is a comprehensive study of all these recent developments. We begin with various multiplexing schemes like SDMA, FDMA, TDMA, CDMA, and OFDM. Then we will discuss the modern trends, challenges, and prospects in satellite communication. We will then delve deep into wireless sensor networks and Internet of Things (IoT) technology.

9.1 INTRODUCTION

In the ever-evolving landscape of communication technology, the quest for more advanced systems is endless. This chapter focuses on the intricacies of Advanced Communication Systems, exploring the cutting-edge methods and technologies, reshaping how we connect and communicate. From the fundamentals of signal processing to the complexities of network architecture, this chapter navigates through a lot of innovations that are driving the future of communication. We discuss topics such as digital multiple access schemes, and error control coding, providing a comprehensive understanding of the principles underpinning advanced communication systems. Furthermore, we examine emerging paradigms like software-defined radio networks and cognitive radio, illuminating their potential to revolutionize communication infrastructure. This chapter introduces the readers to the dynamic world of Advanced Communication Systems.

Some of the latest developments in advanced communication systems design and modeling include the following:

- *5G and Beyond:* The deployment and optimization of 5G networks continue to evolve, with ongoing research focusing on enhancing data rates, reducing latency, and improving energy efficiency. Additionally, exploration into future generations of wireless technology, such as 6G, is underway, aiming to unlock unprecedented speeds and capabilities.
- *Millimeter Wave Communication:* Utilizing higher frequency bands, millimeter-wave communication offers the potential for significantly faster data rates. Research is ongoing to overcome challenges related to propagation, beamforming, and interference mitigation in millimeter-wave systems.

- *Massive MIMO (mMIMO):* Massive Multiple Input Multiple Output (mMIMO) technology, which employs a large number of antennas at both transmitter and receiver, is being further optimized to increase spectral efficiency and enhance network capacity.
- *Software-Defined Radio Networks (SDRs):* SDR architectures allow for flexible and programmable control of network resources, enabling dynamic adaptation to changing communication demands and efficient resource allocation.
- *Cognitive Radio Networks (CRNs):* The cognitive radio network technology is a further development of the software-defined radio concept. In this, there is inherent intelligence ("cognition") of the network, which dynamically optimizes the performance of the radio network at varying channel conditions and user profiles.
- *Network Slicing:* Network slicing facilitates the creation of multiple virtual networks over a shared physical infrastructure, tailored to specific application requirements, such as low latency for autonomous vehicles or high reliability for industrial automation.
- *Quantum Communication:* Quantum communication protocols offer unparalleled security through principles of quantum mechanics, with advancements focusing on extending the range and scalability of quantum key distribution (QKD) systems.
- *Terahertz Communication:* Terahertz frequency bands hold promise for ultra-high-speed wireless communication, with ongoing research addressing challenges related to signal propagation, device integration, and modulation techniques.
- *AI and Machine Learning:* Integration of artificial intelligence and machine learning techniques in communication systems aids in optimizing network performance, predicting user behavior, and automating network management tasks. This will be taken up in the next chapter.

These developments represent just a snapshot of the diverse and dynamic landscape of advanced communication systems design and modeling, with researchers and engineers continually pushing the boundaries of what is possible in the field.

9.2   MULTIPLE ACCESS SCHEMES

Multiple Access schemes allow the simultaneous use of the available spectrum (frequencies allocated for various channels) and other resources such as space and time more effectively. Multiple access schemes can be classified as *reservation based multiple access* (e.g., SDMA, FDMA, TDMA, and CDMA); and *random multiple access*, e.g., ALOHA and CSMA. Depending on the resource that we share, we can have Space Division Multiple Access (SDMA), Time Division Multiple Access (TDMA), Frequency Division Multiple Access (FDMA), and Code Division Multiple Access (CDMA). We will now study each one of them in detail in the following sections.

9.2.1 SPACE DIVISION MULTIPLE ACCESS (SDMA)

Space Division Multiple Access is used in wireless communication systems to increase the capacity and efficiency of data transmission by exploiting spatial resources. It is a form of multiple access technology that enables multiple users to simultaneously transmit and receive data over the same frequency band by using different spatial resources. In traditional wireless communication systems, such as cellular networks, multiple users share the same frequency band, and they are separated using techniques like Time Division Multiple Access (TDMA), Frequency Division Multiple Access (FDMA), or Code Division Multiple Access (CDMA). These techniques allow multiple users to access the same resources by dividing them either in time, frequency, or using different codes.

However, these techniques have their limitations when it comes to increasing capacity and efficiency, especially in dense urban environments where interference and limited spectrum resources are major concerns. SDMA overcomes these limitations by exploiting the spatial dimension. SDMA works by utilizing multiple antennas at both the transmitter and receiver sides. This configuration is commonly known as Multiple Input Multiple Output (MIMO). Each antenna pair can form a unique spatial channel, allowing the system to simultaneously transmit multiple data streams to multiple users.

The basic principle of SDMA is to create orthogonal or nearly orthogonal spatial channels between the transmitter and each receiver. Orthogonal channels ensure that interference between different users is minimized, enabling simultaneous transmission and reception without significant degradation in signal quality. To achieve this, SDMA employs advanced signal processing techniques, such as beamforming and spatial multiplexing. Beamforming involves adjusting the antenna weights and phases to focus the transmitted signals towards the intended users, enhancing the signal strength at the receiver side. Spatial multiplexing takes advantage of the multiple antennas to transmit independent data streams to different users simultaneously.

SDMA also requires accurate channel state information (CSI) of the wireless channel deployed, to enable efficient spatial resource allocation. CSI is obtained through channel estimation techniques, which involve transmitting known pilot signals and measuring the received signals at the receiver side. This information helps in determining the optimal transmission strategy, including beamforming weights and modulation schemes for each spatial channel.

One of the key advantages of SDMA is its ability to increase system capacity and spectral efficiency. By utilizing spatial resources effectively, SDMA enables more users to access the same frequency band simultaneously, thereby increasing the number of supported connections and overall data throughput. Moreover, SDMA can improve the quality of service (QoS) for individual users by reducing interference and enhancing signal strength. This is particularly beneficial in scenarios with high user density, where traditional multiple-access techniques may lead to severe interference and reduced signal quality.

9.2.1.1   Merits and Demerits of SDMA

SDMA offers merits such as increased spectral efficiency by exploiting spatial diversity, enabling higher data rates, and improved network capacity. It also enhances reliability and mitigates interference by spatially separating users.

However, SDMA requires complex antenna systems and signal processing techniques, leading to higher implementation costs. Additionally, it is susceptible to performance degradation in non-line-of-sight scenarios and may face challenges in dense urban environments due to the limited spatial separation feasible. Yet, SDMA remains a promising technology for enhancing the efficiency and performance of wireless communication systems.

9.2.1.2   Applications of SDMA

SDMA is employed in various wireless communication standards, such as Long Term Evolution-Advanced (LTE-Advanced) and 5G (fifth generation) networks, to enhance their performance and support higher data rates. It is expected to play a crucial role in future wireless communication systems as well, including 6G, where even higher capacity and efficiency requirements are anticipated. However, SDMA is not specifically employed in the implementation of Voice Over LTE (VoLTE). Instead, VoLTE relies on Orthogonal Frequency Division Multiplexing (OFDM) and Single Carrier Frequency Division Multiple Access (SC-FDMA) to achieve efficient transmission of voice and data over LTE networks.

Also note that Space Division Multiple Access (SDMA) is not typically used in wired communication systems. SDMA is a technique primarily employed in wireless communication systems to increase spectral efficiency and network capacity by exploiting spatial diversity.

9.2.2   TIME DIVISION MULTIPLE ACCESS (TDMA)

Time Division Multiple Access (TDMA) allows a number of users to access a single radio frequency (RF) channel without interference by allocating unique time slots to each user within each channel. There are two types of TDMA-Narrowband TDMA and Wideband TDMA. In the Wideband TDMA, the entire spectrum is available for each individual user. In TDMA, each user uses the whole channel bandwidth for a fraction of time. In a TDMA system, time is divided into equal time intervals, to be used by each user, called *slots*. Several slots make up a *frame*. *Guard times* are used between the transmission times of each user to minimize crosstalk between channels.

9.2.2.1   Merits and Demerits of TDMA

TDMA permits a flexible bit rate for low bit rate broadcast-type traffic. Also, TDMA allows frame-by-frame monitoring of signal strength or bit error rates to enable mobiles or base stations to initiate and execute handoffs. When TDMA is used exclusively (not with FDMA) provides better bandwidth efficiency, as no guard band is needed between channels. Moreover, TDMA transmits each signal with sufficient guard time between time slots to accommodate time inaccuracies because of clock instability, delay spread, transmission delay due to propagation distance, and the tails of signal pulse due to transient responses.

For mobiles, TDMA on the uplink (mobile to base station link) needs high peak power in transmit mode, which reduces battery life. TDMA also needs a lot of signal processing for matched filtering and correlation detection for synchronization with a time slot. TDMA needs synchronization and if it is lost, the channels may collide with one another. The propagation time for a signal from a mobile station to a base station varies with its distance.

9.2.2.2 Applications of TDMA

TDMA is used in Digital Cellular Systems like GSM, and Digital Advanced Mobile Phone System (D-AMPS) to allow multiple users to share the same frequency band by dividing time into frames and allocating time slots to different users. TDMA is used in Satellite Communication to efficiently utilize the limited bandwidth available for communication between satellites and ground stations. It is also used in Wireless Local Area Networks (WLANs) like Worldwide Interoperability for Microwave Access (WiMAX) to allow multiple users to access the same wireless channel by dividing time into frames and allocating time slots to different users. TDMA is also used in digital broadcasting systems to transmit multiple digital TV or radio channels within the same frequency band by dividing time into slots and allocating them to different channels; and for industrial communication systems, such as Supervisory Control and Data Acquisition (SCADA) systems, where multiple sensors and devices need to transmit data over a shared communication medium.

9.2.3 FREQUENCY DIVISION MULTIPLE ACCESS (FDMA)

FDMA enables multiple users to share an entire communication resource, the available frequency spectrum. In FDMA, each user uses a different band of frequencies to communicate the information, pertaining to a channel. Multiple users are isolated using bandpass filters. Frequency guard bands are provided between adjacent channels to minimize crosstalk.

FDMA allows multiple users to share the available bandwidth of a communication channel. In FDMA, the frequency spectrum is divided into multiple non-overlapping frequency bands, with each band allocated to a different user or communication channel. Each user is assigned a specific frequency band within which they can transmit data independently of other users. FDMA is based on the principle of frequency separation, enabling simultaneous communication by different users without interference. Each user is assigned a unique frequency band, and they can transmit data within their allocated band without affecting other users sharing the same channel.

9.2.3.1 Merits and Demerits of FDMA

One of the primary merits of FDMA is its simplicity, as it does not require complex synchronization mechanisms compared to other multiple access techniques. Additionally, FDMA allows for efficient use of bandwidth by dividing it into distinct frequency bands, which can be dynamically allocated based on user demand.

However, FDMA also has limitations. It is not as spectrally efficient as other multiple access techniques like Code Division Multiple Access (CDMA) or Orthogonal Frequency Division Multiple Access (OFDMA), since it requires guard bands to prevent interference between adjacent frequency bands. This can result in some spectrum wastage, especially in use cases with a large number of users or varying data rates.

Despite these limitations, FDMA remains a fundamental multiple-access technique, particularly in legacy communication systems such as analog cellular networks and satellite communication systems. It provides a straightforward and effective means of sharing communication channels among multiple users, making it a cornerstone of telecommunications technology.

9.2.4   SPECTRAL EFFICIENCY

The efficient use of available spectrum is the most desirable feature of a mobile communication system. Spectral efficiency can be increased by reducing the channel bandwidth, information compression (by low bit rate speech coding), using variable bit rate CODECs (Coder-Decoder), and enhanced channel assignment algorithms. Spectral efficiency of a mobile communication system depends on the modulation scheme used and the multiple access scheme.

9.2.4.1   Spectral Efficiency of Modulation

The *spectral efficiency of modulation* is defined as:

$$\eta_m = \frac{Total\,Number\,of\,Channels\,Available\,in\,the\,System}{Bandwidth \times Total\,Coverage\,Area}$$

$$= \frac{\frac{B_w}{B_c} \times \frac{N_c}{N}}{B_w \times N_c \times A_c} \tag{9.2.1}$$

$$= \frac{1}{B_c \times N \times A_c}\,Channels/MHz/km^2 \tag{9.2.2}$$

where: η_m is the modulation efficiency (in channels/MHz/km^2); B_w is the bandwidth of the system (in MHz); B_c is the channel spacing (in MHz); N_c is the total number of cells in the covered area; N is the frequency reuse factor of the system (cluster size); and A_c is the area covered by a cell (in km^2). As per the above equation, spectral efficiency of modulation can be increased by reducing the channel spacing (B_c) and the frequency reuse factor (N).

Another definition of spectral efficiency of modulation is *Erlangs/MHz/km^2*. It is given as:

$$\eta_m = \frac{Maximum\,Total\,Traffic\,Carried\,by\,the\,System}{System\,Bandwidth \times Total\,Coverage\,Area}$$

$$= \frac{Total\,Traffic\,Carried\,by\,\left(\frac{B_w/B_c}{N}\right)\,Channels}{B_w \times A_c} \tag{9.2.3}$$

By introducing the trunking efficiency factor, η_t, in equation 9.2.3; which is less than 1 and a function of the blocking probability and the number of available channels per cell; the spectrum efficiency of modulation is given as:

$$\eta_m = \frac{\eta_t \times \left(\frac{B_w/B_c}{N} \right)}{B_w \times A_c} \qquad (9.2.4)$$

$$= \frac{\eta_t}{B_c N A_c} \qquad (9.2.5)$$

where: η_t is a function of the blocking probability and the total number of available channels per cell, $\left[\frac{B_w/B_c}{N} \right]$.

Based on equation 9.2.5, we can conclude that:

- The voice/data quality will depend on the frequency reuse factor, N, which is a function of the signal-to-interference ratio (SIR) of the modulation scheme used in the mobile communication system.
- The relationship between the system bandwidth, B_w, and the amount of traffic carried by the system is nonlinear, i.e., for a given percentage increase in B_w, the increase in the traffic carried by the system is more than the increase in B_w.
- From the average traffic per user (*Erlang/user*)during the busy hour and *Erlang/MHz/km²*, the capacity of the system in terms of *users/km²/MHz* can be calculated. *The spectral efficiency of modulation is dependent on the blocking probability.*

9.2.4.2 Multiple Access Spectral Efficiency

It is the ratio of the total time or frequency dedicated for the transmission of traffic to the total time or frequency available to the system. Hence, the multiple access spectral efficiency is a dimensionless number which is having an upper limit of unity. In FDMA, the multiple access spectral efficiency is reduced because of the guard bands between the channels and also due to the *signaling channels*. In TDMA, the multiple access spectral efficiency is reduced because of guard time and the synchronization sequence.

For the wideband TDMA, the multiple access spectral efficiency is given as:

$$\eta_a = \frac{\tau M_t}{T_f} \qquad (9.2.6)$$

where: τ is the duration of a time slot that carries data, M_t is the number of time slots per frame, and T_f is the duration of each frame. In equation 9.2.6, it is assumed that the total available bandwidth is shared by all users. For the narrowband TDMA schemes, the total band is divided into a number of sub-bands, each using the TDMA technique. For the narrowband TDMA system, frequency domain efficiency is not unity as the individual user channel does not use the whole frequency band available to the system. The multiple access spectral efficiency of the narrowband TDMA system is given as:

$$\eta_a = \left(\frac{\tau M_t}{T_f}\right)\left(\frac{B_u N_u}{B_w}\right) \tag{9.2.7}$$

where: B_u is the bandwidth of an individual user during the allotted time slot, N_u is the number of users sharing the same time slot in the system, but having access to different frequency sub-bands, and B_w is the total allocated bandwidth of the TDMA system.

In the case of FDMA, multiple access spectral efficiency is given as:

$$\eta_a = \frac{B_c N_T}{B_w} \leq 1 \tag{9.2.8}$$

where: η_a is the multiple access spectral efficiency; B_c is the channel spacing; N_T is the total number of traffic channels in the coverage area; and B_w is the bandwidth of the system.

9.2.4.3   Overall Spectral Efficiency

The overall spectral efficiency, η, is the product of both the modulation and multiple access spectral efficiencies. Hence:

$$\eta = \eta_m \times \eta_a \tag{9.2.9}$$

9.2.5   CODE DIVISION MULTIPLE ACCESS (CDMA)

In CDMA, different channels or users share the entire spectrum of frequencies allocated, but each user is assigned a particular code for conveying information. CDMA is basically a digital communication technology. Each user employs a particular code to convey a digital one and digital zero. Code Division Multiple Access (CDMA) is a spread spectrum multiple access scheme widely used in modern telecommunications systems, particularly in wireless communication networks such as cellular systems. Unlike FDMA or TDMA, which allocate distinct frequency bands or time slots to different users, CDMA allows multiple users to transmit simultaneously over the same frequency band by using unique spreading codes. CDMA was developed by Qualcomm, Inc. and standardized by the Telecommunications Industry Association (TIA) as an Interim Standard IS-95. This system supports a variable number of users in 1.25MHz wide channels using Direct Sequence Spread Spectrum (DSSS).

The essence of CDMA lies on the concept of spreading codes, also known as pseudorandom noise (PN) codes or spreading sequences. These codes are used to modulate the data from each user before transmission. Each user is assigned a unique spreading code, which spreads the narrowband signal across a much wider bandwidth. The spreading codes are carefully designed to have low cross-correlation with one another, allowing multiple users to share the same frequency band without significant interference.

A main advantage of CDMA is its ability to combat interference and multipath fading. Because signals from each user are spread across a wide bandwidth, it appears as noise to other users not sharing the same spreading code. As a result, CDMA inherently provides a degree of resistance to narrowband interference and allows for coexistence of multiple users in the same frequency band. Moreover, CDMA offers inherent security benefits due to the pseudo-random nature of the spreading codes. Since the signal appears as noise to users not employing the correct spreading code, it is inherently more difficult for unauthorized users to eavesdrop or intercept the communications.

CDMA is widely used in wireless cellular communication systems, particularly in 3G and 4G networks. In CDMA-based cellular networks, each user is allocated a unique spreading code, and all users share the same frequency band simultaneously. This allows for more efficient use of the available spectrum and supports a larger number of users compared to FDMA or TDMA-based systems. In addition to that, CDMA has been a key enabling technology for high-speed data transmission in wireless networks, as it facilitates the use of advanced modulation and coding techniques such as multi-carrier modulation and turbo coding.

To summarize, CDMA is a robust and versatile multiple-access technique that has played a crucial role in the evolution of wireless communication systems. Its ability to provide efficient spectrum utilization, resistance to interference, and inherent security features make it a prominent candidate of modern telecommunications technology.

9.2.5.1   CDMA2000 and W-CDMA

CDMA2000 and Wideband Code Division Multiple Access (W-CDMA) are both standards for wireless communication technologies, but they are based on different underlying principles and are used in different types of networks. The CDMA2000 (a.k.a. 3G Partnership Project 2, 3GPP2) is based on Code Division Multiple Access (CDMA) technology and is primarily used in CDMA-based cellular networks, such as those deployed by carriers like Verizon and Sprint in the US. W-CDMA (Wideband Code Division Multiple Access), also known as Universal Mobile Telecommunications System (UMTS), is based on CDMA technology as well but is standardized by the 3rd Generation Partnership Project (3GPP) and is used in GSM-based cellular networks worldwide, including carriers like AT&T and T-Mobile in the US.

CDMA2000 operates primarily in the 800 MHz and 1900 MHz frequency bands, which are used by CDMA carriers in regions like North America, South Korea, and parts of Latin America. W-CDMA operates in various frequency bands, including the 850 MHz, 900 MHz, 1700/2100 MHz (Advanced Wireless Services), and 1900 MHz bands, depending on the region and carrier deployment.

Note that CDMA2000 networks are generally not backward compatible with GSM networks, requiring separate infrastructure and devices. W-CDMA networks are designed to be backward compatible with existing GSM networks, allowing for a smoother transition from 2G to 3G services and interoperability with GSM-based devices.

CDMA channels can be modeled in MATLAB. Refer MATLAB Communications Toolbox documentation for details.

9.3   ORTHOGONAL FREQUENCY DIVISION MULTIPLEXING (OFDM)

Orthogonal Frequency Division Multiplexing (OFDM) is a widely used modulation technique in modern wireless communication systems, including WiFi, 4G LTE, and 5G networks. OFDM divides the available frequency spectrum into multiple narrowband subcarriers, each spaced orthogonally to one another. These subcarriers are closely packed together, allowing efficient use of the available spectrum.

A major advantage of OFDM is its ability to mitigate the effects of multipath propagation, a common issue in wireless communication where signals travel along multiple paths due to reflections, diffraction, and scattering. OFDM accomplishes this by using a guard interval between symbols, which helps prevent intersymbol interference caused by delayed signal arrivals. OFDM also offers high spectral efficiency by packing multiple subcarriers within the available bandwidth. Each subcarrier can be modulated independently, allowing for flexible adaptation to varying channel conditions. Additionally, OFDM supports adaptive modulation and coding, enabling the system to dynamically adjust data rates to maximize throughput and reliability. Moreover, OFDM is robust against narrowband interference and frequency-selective fading, making it suitable for high-speed data transmission in challenging wireless environments. It also supports multiple access techniques such as Orthogonal Frequency Division Multiple Access (OFDMA), which further enhances its efficiency in multi-user applications.

Summarizing, OFDM is a versatile and efficient modulation technique that has become the foundation of modern wireless communication systems. Its ability to address multipath propagation, achieve high spectral efficiency, and support flexible adaptation to channel conditions make it indispensable in delivering high-speed and reliable wireless connectivity.

9.4   RANDOM MULTIPLE ACCESS SCHEMES

Random Multiple Access schemes refer to communication protocols used in computer networks. In it, each device has an equal privilege to access the network and transmit data. These protocols allow nodes to send and receive data without taking turns or waiting for permission. The earliest scheme of random access is the ALOHA system. In an ALOHA (a.k.a. pure ALOHA) system, each transmitter begins to transmit whenever it has data to transmit. If there is a collision, each active transmitter randomly chooses a backoff time and retransmits the data packet after the backoff time. In pure ALOHA, stations transmit data packets whenever they have them ready, without waiting for any specific timing. This leads to the possibility of collisions if two or more stations transmit simultaneously. In slotted ALOHA, time is divided into fixed-size slots, and stations are required to transmit only at the beginning of each time slot. This synchronized timing reduces the chances of collisions occurring.

In slotted ALOHA, collisions are easier to detect because they happen at pre-defined intervals, simplifying the retransmission process. Slotted ALOHA is more efficient than pure ALOHA in terms of channel utilization and throughput. By synchronizing transmissions to time slots, slotted ALOHA reduces the probability of collisions and increases the overall efficiency of the system.

9.4.1 CARRIER SENSE MULTIPLE ACCESS (CSMA)

Carrier Sense Multiple Access with Collision Detection (CSMA/CD) and Carrier Sense Multiple Access with Collision Avoidance (CSMA/CA) are two variants of the Carrier Sense Multiple Access (CSMA) protocol used in network communication.

CSMA/CD is primarily used in Ethernet LANs (Local Area Networks) to manage access to the shared communication medium. Before transmitting, the nodes listen to the medium to check if it is idle (carrier sensing). If the medium is busy, they wait for it to become idle before transmitting. If two or more nodes transmit simultaneously, a collision occurs. CSMA/CD nodes detect collisions by listening to the medium while transmitting. Upon detecting a collision, they stop transmission immediately and send a jam signal to alert other nodes of the collision. After a collision, nodes wait for a random backoff period before attempting to retransmit their packets. This random backoff helps reduce the likelihood of collisions reoccurring. CSMA/CD is commonly used in Ethernet networks, particularly in wired LANs, to ensure fair and efficient access to the shared transmission medium.

CSMA/CA is primarily used in wireless LANs (WiFi) to manage access to the shared wireless channel. Like CSMA/CD, nodes using CSMA/CA listen to the medium before transmitting to avoid collisions. However, due to the hidden terminal problem in wireless networks, where nodes may not be able to detect transmissions from other stations, CSMA/CA employs a collision avoidance mechanism rather than collision detection. Before transmitting, nodes using CSMA/CA protocol wait for a random backoff period, which helps reduce the likelihood of collisions with other stations. Additionally, they use techniques like Request to Send (RTS) and Clear to Send (CTS) to reserve the channel and avoid collisions with hidden terminals.

CSMA/CA helps improve the efficiency of wireless LANs by reducing the chances of collisions and maximizing the utilization of the shared wireless channel.

9.5 ERROR CONTROL CODING

Error control coding involves adding extra bits to the data bits being conveyed from the transmitter, to detect possible errors introduced at the channel, at the receiver side. By suitable decoding techniques at the receiver, the errors can be detected and corrected at the receiver. However, adding error control coding to the message being transmitted increases the bandwidth occupancy; and this in turn reduces the bandwidth efficiency; but at the same time reduces the required received power for a given bit error rate. Thus error control coding trades-off bandwidth efficiency with power efficiency. The higher-level modulation schemes (M-ary keying) decrease bandwidth occupancy but increase the required received power, and hence trade-off power efficiency with bandwidth efficiency. Note that the bandwidth efficiency η_B; if the transmitted data rate R bits per second, and B is the bandwidth occupied by the modulated RF signal; is expressed as:

$$\eta_B = \frac{R}{B} \ bps/Hz \tag{9.5.10}$$

Bandwidth efficiency shows how efficiently the allocated bandwidth is utilized and is the ratio of the *throughput data rate per Hertz* in a given bandwidth. The system capacity of a digital mobile communication system is directly related to the bandwidth efficiency of the modulation scheme, since a modulation with a greater value of bandwidth efficiency η_B will be able to transmit more data in a given spectrum allocation. The upper bound on the achievable bandwidth efficiency is given by *Shannon's Channel Coding theorem*. The theorem states that for an arbitrarily small probability of error, the maximum possible bandwidth efficiency is limited by the noise in the channel, and given by by the channel capacity formula. Note that the Shannon's bound is applicable for AWGN non-fading channels:

$$\eta_{B\,max} = \frac{C}{B} = \log_2\left(1 + \frac{S}{N}\right) \tag{9.5.11}$$

where C is the channel capacity (in bps), B is the RF bandwidth (in Hertz), and $\frac{S}{N}$ is the signal-to-noise ratio at the receiver.

9.5.1 CHANNEL CODING

Channel coding protects the digital data being transmitted from errors by selectively introducing redundant bits in the transmitted data. Some of the codes used in channel coding can only detect errors in received data and they are known as *error detection codes*; where as some of the codes can detect and correct errors and are known as *error correction codes*.

The Shannon's Channel Capacity formula can also be stated as:

$$C = B\log_2\left(1 + \frac{P}{N_0 B}\right) \tag{9.5.12}$$

where C is the channel capacity (bits per second), B is the transmission bandwidth (Hz), P is the received signal power (Watts), and N_0 is the single-sided noise power density (Watts/Hz). We can write the received power P as $P = E_b R_b$, where E_b is the average bit energy, and R_b is the transmission bit rate. Thus we can write the bandwidth efficiency as:

$$\frac{C}{B} = \log_2\left(1 + \frac{E_b R_b}{N_0 B}\right) \tag{9.5.13}$$

The purpose of error detection and error correction coding is to introduce redundancies in the transmitted data to improve the wireless link performance. The redundant bits increase the raw data rate in the link, and hence it increases the bandwidth requirements for a fixed source data rate. This in turn reduces the bandwidth efficiency of the link in high SNR conditions, but gives excellent BER performance at low SNR values.

9.5.2 LINEAR BLOCK CODES

Linear block codes are *forward error correction (FEC)* codes that enable a limited number of errors to be detected and corrected without retransmission. In block codes, k information bits are encoded into n code bits by adding $n - k$ redundant bits, the block code is called a (n,k) code. The rate of the code is defined as $R_c = k/n$ and is equal to the rate of information divided by the raw channel rate. The two important parameters of codes are the distance between two codewords (which is the number of elements in which the two codewords differ) and the weight of the codeword (which is the total number of non-zero elements). Now we will examine some examples for block codes.

9.5.2.1 Hamming Codes

A binary (n,k) Hamming code has the property that:

$$(n,k) = (2^m - 1, 2^m - 1 - m)$$

where k is the number of information bits used to form a n-bit codeword, and m is any positive number, greater than 1. The number of parity bits is $n - k = m$. It is known that a (n,k) Hamming code can detect up to $\lfloor (n-k+1)/2 \rfloor$ errors and correct up to $\lfloor (n-k)/2 \rfloor$ errors.

9.5.2.2 Hadamard Codes

Hadamard codes are obtained by selecting as codewords the rows of a *Hadamard matrix*. A Hadamard matrix A is an $N \times N$ matrix of 1s and 0s, such that each row differs from any other row in exactly $N/2$ positions. One row contains all zeros with the remaining contain $N/2$ zeros and ones. The minimum distance for these codes is $N/2$. For $N = 2$, the Hadamard matrix A is

$$A = \begin{bmatrix} 0 & 0 \\ 0 & 1 \end{bmatrix}$$

When $N = 2^m$ (m being a positive integer), Hadamard codes of other lengths are possible, but the codes are *not linear*.

9.5.2.3 Golay Codes

The Golay codes are linear binary $(23, 12)$ codes with a minimum distance of and an error correction capability of 3. They are the only nontrivial perfect codes. This means that any combination of errors that falls within the code's capability to detect will be identified as an error. For error-correcting codes, a perfect code can both detect and correct all possible patterns of a certain number of errors within the code.

This means that not only will the code identify the presence of errors, but it will also be able to precisely locate and correct those errors, restoring the original information.[1]

9.5.2.4 Cyclic Codes

Cyclic codes are a subset of linear codes, in which the cyclic shift of a codeword results in another codeword. A cyclic code can be formed by using a generator polynomial $g(p)$ of degree $(n-k)$. The generator polynomial of an $(n-k)$ cyclic code is a factor of $p^n + 1$ and has the general form:

$$g(p) = p^{n-k} + g_{n-k-1}p^{n-k-1} + \ldots + g_1 p + 1$$

A message polynomial can also be defined as

$$x(p) = x_{k-1}p^{k-1} + \ldots + x_1 p + x_0$$

where $(x_{k-1}, \ldots, x_0)$ represents the k information bits. The resulting codeword $c(p)$ can be written a

$$c(p) = x(p)g(p)$$

where $c(p)$ is a polynomial of degree less than n. Encoding is performed by a linear feedback shift register based on either the generator or parity polynomial.

9.5.2.5 BCH Codes

Bose-Chaudhuri-Hocquenghem (BCH) codes are a class of cyclic error-correcting codes that provide both error detection and error correction capabilities. They were developed by Raj Bose and D. K. Ray-Chaudhuri in 1960 and independently by Claude Berrou and Alain Glavieux in 1986. BCH codes are a subclass of cyclic codes. The block length of the codes is $n = 2^m - 1$ for $m \geq 3$, and the number of errors that can be corrected is bounded by $t < (2^m - 1)/2$.

9.5.2.6 Reed-Solomon Codes

Reed-Solomon codes are nonbinary codes which are capable of correcting burst errors. The block length of the codes is $n = 2^m - 1$. These can be extended to 2^m or $2^m + 1$. The number of parity symbols that must be used to correct e errors is $n - k = 2e$. The minimum distance is $d_{min} = 2e + 1$, which is the largest possible of any linear code. The $(63, 47)$ Reed-Solomon code used in US Cellular Digital Packet Data (CDPD) uses $m = 6$ bits per code symbol.

---

[1] Hamming codes are also perfect.

9.5.3 CONVOLUTIONAL CODES

A convolutional code is generated by passing the information sequence through a finite state shift register. In general, the shift register contains N $k - bit$ stages and m linear algebraic function generators based on the generator polynomials. The input data is shifted into and along the shift register, k bits at a time. The number of output bits for each k-bit user input data sequence is n bits. The code rate is $R_c = k/n$. The parameter N is called the *constraint length* and is the number of input data bits that the current output is dependent upon.

The convolutional codes can be decoded using the *Viterbi Algorithm*, Fano's Sequential Decoding, the Stack Algorithm or by Feedback decoding.

9.5.4 TRELLIS CODED MODULATION (TCM)

It combines both coding and modulation to achieve significant coding gains without compromising bandwidth efficiency. TCM employs redundant nonbinary modulation along with a finite state encoder that decides the selection of modulation signals to generate coded signal sequences. With TCM, coding gains as large as 6dB can be obtained without any bandwidth expansion or reduction in the effective information rate.

9.5.5 TURBO CODES

Turbo codes combine the capabilities of convolutional codes with channel estimation theory, and can be considered as parallel convolutional codes. When implemented properly, turbo codes allow coding gains which are far superior to all previously discussed error-correcting codes, and allow a wireless communication link to come close to realizing the Shannon capacity bound. Turbo codes are used in 3G wireless standards.

MATLAB Communications Toolbox along with Simulink contain a number system building blocks to simulate and implement Turbo coding.

9.6 SOFTWARE DEFINED RADIO NETWORKS

The credit for the conceptualization and implementation of the world's first Software Defined Radio (SDR) system goes to Joseph Mitola III, who introduced the concept of SDR in his doctoral dissertation at KTH Royal Institute of Technology in Sweden in 1995. His work laid the foundation for the development of SDR technology, which has since revolutionized the field of wireless communications. The SDR Forum defines SDRs as a "Radio that provides software control of a variety of modulation techniques, wide and narrow band operation operation, communication security functions and waveform requirements of current and evolving standards over a broad frequency range".

The Software Defined Radio refers to a radio communication system where the typical hardware building blocks such as mixers, filters, amplifiers, modulators/ demodulators, and detectors are implemented using software running on a computer or embedded system. Note that in a typical,conventional radio communication system all the above-mentioned subsystems are implemented in hardware. The main advantage of SDR technology is the flexibility, and the ease it allows for the reconfiguration of radio parameters and protocols; through software updates, rather than requiring hardware changes or upgrades. One of the main lacunae of conventional radio systems is that two groups of users having dissimilar different types of radio hardware system are not able to communicate with each other due incompatibility. This lacunae can be effectively solved using Software Defined Radio.

The SDR is a wireless communication device that works with any type of communication system; such as mobile phone, a WiFi transceiver, or a pager or a AM or FM radio, or even a satellite communication device. In SDR, the hardware and software building blocks are implemented based on digital signal processing techniques. For implementing highly complex SDR applications, the sampling rate at the different stages cannot be uniform (same) and hence *multi-rate digital signal processing* is used.

The SDR can be considered as an *open architecture system* that builds up a communication platform by modular, standardized flexible hardware blocks. The SDR supports a multitude of wireless communication services in a single design.

The SDR comprises of two major parts:

1. Hardware subsystem
2. Software subsystem

If we look at the hardware subsystem, the main building blocks are the following:

- Intelligent (Smart) Reconfigurable Antenna.
- Programmable RF Module
- High-performance DAC and ADC
- The Interconnects

9.6.1 INTELLIGENT, SMART, AND RECONFIGURABLE ANTENNA

SDR antennas are based on several separate microstrip antennas to cover a broad range of frequencies. The ideal antenna used in SDR is a self-adaptable, self-aligned, and self-healing microstrip patch antenna which is capable of complete adaptation to the intended use case and the transmission environments. Micro Electro Mechanical System (MEMS) antenna is a promising candidate for broadband reconfigurable antenna used in SDRs. By using MEMS switches, the desired antenna slot elements are switched in and out, for a new frequency band.

9.6.2 PROGRAMMABLE RF MODULE

In the SDRs, a bank of RF modules is used to cover the entire frequency band. Again, thanks to compact, low-loss MEMS devices, the implementation of high-performance RF devices with a high level of integration of circuits is possible.

MEMS technology enhances the performances and flexibility of several RF components including low-phase noise voltage-controlled oscillators (VCO) using high Q-resonators; wideband varactors and phase shifters using variable capacitors and switched capacitor networks; and tunable filters by using MEMS-based variable reactive elements and switches.

9.6.3   DAC AND ADC MODULES

The DAC and ADC have the unique task of conversion between digital to analog and vice-versa, and they proceed the RF module. Considerable performance improvement wrt to the flexibility of SDR can be achieved by pushing the DAC/ADC closer to the smart antenna. Conventional electronic DACs/ADCs are moving the envelop to achieve to obtain more resolution and faster conversion rates. However, employing superconductor (RSFQ), and optical converters; even higher performance can be achieved.[2]

9.6.4   DIGITAL SIGNAL PROCESSING

The main enabler for SDR implementation is digital signal processors and processing. The capability of DSPs in signal compression, noise cancellation, multidimensional filtering, Adaptive processing, signal detection and estimation, and array processing make them ideal for use in SDRs. Among several DSP techniques used in SDR platforms; sampling technique, rate conversion, and multirate processing have been instrumental.

9.6.5   INTERCONNECTS IN SDR

The following crucial issues must be addressed for an interconnect strategy to be successful:

- Open standards
- Handling various communication protocols
- Fulfilling the growing demands for throughput and speed.
- Making a connection to conventional circuit networks

Bus architecture, switch fabric architecture, and tree architecture are the three primary interconnect architectures. It should be highlighted, that the use of switch fabric architecture, which is the most commonly used in SDR, necessitates stringent requirements and performance in terms of throughput and packet latency on the switch fabric.

Switch fabric architecture allows for flexible and efficient routing of data between different components or modules within the SDR. It typically involves a network of

---

[2] Rapid Single-Flux-Quantum (RSFQ) logic is a superconductor IC technology that, with only a modest number of researchers worldwide, has produced some of the world's highest-performance digital and mixed-signal circuits.

switches that can dynamically connect different input and output ports to facilitate communication between various processing elements such as ADCs/DACs, digital signal processors, and general-purpose processors. This flexibility is well-suited for the dynamic and adaptive nature of SDRs, where different signal processing tasks may require varying levels of computational resources and data throughput.

9.6.6   SOFTWARE SUBSYSTEMS IN SDR

The SDR needs a fast and efficient method for generating and verifying the required signal-processing algorithm to configure a subsystem within it. Software subsystems in SDRs include signal processing modules for tasks such as modulation/demodulation, filtering, channel equalization, error correction coding/decoding, and digital beamforming; radio control for RF front end, ADCs/DACs, and other RF components; protocol stacks to support multiple communication standards and protocols, such as WiFi, Bluetooth, LTE, GSM, etc; Operating System (OS) Interface which provides an interface between the SDR hardware and the higher-level software applications. It manages tasks such as memory allocation, task scheduling, interrupt handling, and device drivers. The software subsystem also take care of networking that facilitate communication between multiple SDR nodes or between SDRs and external systems. This may involve protocols like TCP/IP, UDP/IP, or custom networking protocols.

The user interfaces do the configuration, monitoring, and control; and they may be command-line interfaces (CLI), graphical user interfaces (GUI), or web-based interfaces. The firmware and bootloader in software subsystem manage the low-level software that initializes the SDR hardware and loads the main software components during boot-up. Finally, in SDRs used for military, public safety, or other sensitive applications, the security module handle tasks such as encryption/decryption, authentication, access control, and secure communications.

9.7   COGNITIVE RADIO NETWORKS (CRN)

In recent years, cognitive radio networks have gained considerable research and development attention from the engineers and researchers working in wireless communication systems. The cognitive radio technology is developed to enhance the radio resource usage of the wireless networks. Cognitive radio networks (CRNs) are a revolutionary development in wireless communication systems. They are designed to address the inefficiencies and limitations of traditional static spectrum allocation methods by means of *dynamic spectrum access*. This technology promises to revolutionize the way we utilize the radio frequency (RF) spectrum, making it more efficient and flexible. The cognitive radio technology is a further development of the software-defined radio concept, which was originated in the 20th century. In early 2000s, cognitive radio research gained momentum with an emphasis on giving radio systems intelligence ("cognition") and the ability to make decisions on their own. CRNs seek to enhance overall system performance and spectrum usage by dynamically adjusting radio characteristics and behaviors in response to user needs, regulatory restrictions, and environmental conditions.

The back bone of cognitive radio networks is the cognitive radios, which are intelligent devices capable of autonomously sensing their environment, adapting their parameters, and making decisions to optimize spectrum utilization while avoiding interference. These radios possess advanced sensing, learning, and decision-making capabilities, allowing them to identify unused spectrum bands (known as white spaces in wireless communication parlance) and efficiently use them without causing adverse interference to the licensed users.

CRNs operate on the principle of spectrum sharing, where cognitive radios access spectrum bands that are not being used by the primary (licensed) users. This dynamic spectrum access enables efficient utilization of the allocated RF spectrum, which is a scarce and expensive resource. By leveraging underutilized spectrum bands, CRNs can significantly increase spectral efficiency and alleviate the problem of spectrum scarcity, especially in densely populated urban areas where spectrum congestion is a common issue. One of the key merits of cognitive radio networks is their ability to improve spectrum utilization and increase network capacity without requiring additional spectrum allocation. This is particularly important as the demand for wireless communication continues to grow exponentially, driven by emerging technologies such as the Internet of Things (IoT), 5G and 6G, Wireless Sensor Networks, and so on. CRNs have the potential to unlock new opportunities for spectrum sharing and enable the seamless coexistence of diverse wireless systems and applications. The cognitive radio networks offer the following benefits:

- *Dynamic Spectrum Access:* Cognitive radios can dynamically adapt their transmission parameters (such as frequency, power, and modulation) based on real-time spectrum availability and user requirements, maximizing spectral efficiency.
- *Interference Mitigation:* CRNs employ sophisticated interference mitigation techniques to ensure that secondary users do not cause interference to primary users. This includes spectrum sensing, and power control mechanisms to detect and avoid occupied spectrum bands.
- *Improved Reliability and Resilience:* Cognitive radio networks can drastically improve the reliability and resilience of wireless communication systems by dynamically reconfiguring network parameters in response to changing environmental conditions, spectrum availability, and network dynamics.
- *Spectrum Efficiency:* By efficiently accessing underutilized bands of spectrum, CRNs can achieve higher spectral efficiency compared to traditional static spectrum allocation methods, leading to improved network performance and user experience.
- *Support for Heterogeneous Devices and Applications:* Cognitive radio networks can support a wide range of devices and applications with diverse quality-of-service (QoS) requirements. This includes applications requiring high bandwidth, mission-critical communications, and low-power IoT devices.

Cognitive radio networks do, however, also have a number of demerits, such as complicated spectrum sensing and management, problems with regulations and policies, security and privacy issues, and interoperability issues. It will be imperative to overcome these obstacles before cognitive radio networks can be widely implemented and used in practical applications. In conclusion, cognitive radio networks represent a revolutionary technology with the potential to completely change wireless communication systems through the enhancement of spectral efficiency, the facilitation of dynamic spectrum access, and the ability to meet the always-growing demand for wireless services. CRNs are anticipated to have a significant impact on the advancement of wireless communications through continued research and development, creating new domains for innovation and societal impact.

9.8 NEW TRENDS IN SATELLITE COMMUNICATIONS

Satellite communication technology is used in almost all forms of communication. NR (New Radio) technology is made possible by 5G integration with satellite communication, which helps to prepare the way and promotes the development of future telecommunications techniques. There are many new developments and trends in satellite communications, particularly with regard to Low Earth Orbit constellations, which have the potential to significantly lower latency when compared to Geosynchronous Orbit Satellites. These constellations are crucial to the advancement of modern telecommunications and technology toward the Interconnection of Everything (IoE).

Non-geostationary orbit (NGSO) satellites, which offer exciting new communication capabilities to provide non-terrestrial connectivity solutions and to support a wide range of digital technologies from various industries, are defining the next phase of satellite technology. In comparison to traditional geo-stationary orbit (GSO) satellites, NGSO communication systems are recognized for a number of important characteristics, including reduced propagation delays, smaller sizes, and lower signal losses. NGSOs may also make it possible to provide latency-critical applications via satellites. NGSO aims to address the primary barriers to commercializing GSO satellites for wider use by offering a significant increase in communication speed and energy efficiency.

The smooth integration with GSO satellites is the primary obstacle to the adoption of NGSO satellite-based communication systems. The coexistence with GSO systems in terms of spectrum access and regulatory issues, satellite constellation and architecture designs, resource management problems, and user equipment also pose challenges in implementation.

The recent advancements in satellite communication include massive use and dependence on Low Earth Orbit (LEO) satellite constellations to reduce the latency manifold, compared to Geosynchronous Orbit Satellite and is a technology enabler for Interconnection of Everything (IoE). Another development is use of optical wireless communication for inter-satellite links using LASER (for enhanced anti-interference, because of its narrow beam property) and ground-to-satellite links. New Radio (NR) is the recent development in 5G wireless communication standard, which has the key features like:

- Higher Data Rates (better than previous mobile networks like 4G LTE), through the use of wider bandwidths, advanced modulation schemes, and more efficient encoding techniques.
- Lower Latency - The New Radio is designed for ultra-low latency communication, which is crucial for applications like autonomous vehicles-autonomous driving and vehicle-to-vehicle (V2V) communication-(typical latency is 1-10ms or even $1\text{-}10\mu s$), industrial automation ($100\mu s$ to 1ms), and augmented reality (10-20ms).
- Massive MIMO (mMIMO)- New Radio deploys massive MIMO technology, which uses a large number of antennas at the base station to communicate with multiple users simultaneously, to increase network capacity, and to improve spectral efficiency and coverage.
- Beamforming: New Radio supports advanced beamforming techniques to enhance signal strength, coverage, and spectral efficiency, especially in environments with high interference.
- Flexible Numerology and Frame Structure: This allows operators to adapt the system parameters based on specific use cases, to ensure efficient spectrum utilization and to support diverse requirements across different applications.
- Network Slicing: NR enables network slicing, which allows operators to partition their network into multiple virtual networks, each tailored to serve specific types of services or customers. This enables efficient resource allocation, isolation, and customization of services based on needs.
- Enhanced Mobile Broadband (eMBB): New Radio is optimized to deliver enhanced mobile broadband services with higher throughput, improved coverage, and better user experience for applications such as high-definition video streaming, online gaming, and virtual reality.
- Ultra-Reliable Low Latency Communication (URLLC): New Radio provides support for ultra-reliable low latency communication, catering to applications that demand high reliability and real-time response, like industrial automation, mission-critical communications, and remote surgery.
- Massive IoT (Internet of Things): New Radio supports massive IoT deployments, allowing a large number of IoT devices to connect simultaneously, with low power consumption, extended coverage, and efficient resource utilization. This enables various IoT applications like smart cities, agriculture, health care, and many more.

9.9   WIRELESS SENSOR NETWORKS (WSN)

Wireless Sensor Network (WSN) is an infrastructure-less wireless network that is deployed in a large number of wireless sensors in an ad-hoc manner that is used to monitor the system, physical or environmental conditions. Sensor nodes used in WSN have sensors for monitoring conditions like temperature, humidity, or water

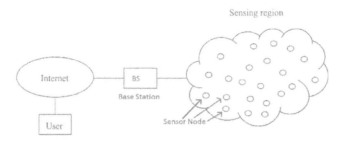

Figure 9.1 Block Diagram of a Wireless Sensor Network

content in the soil etc., with the onboard processor that manages and monitors the environment in a particular area. They are connected to the Base Station which acts as a processing unit in the WSN System. In turn, the Base Station (BS) in a wireless sensor network is connected through the Internet to share data, and to receive control and reconfigurations commands. The security goals, like data confidentiality, data integrity, data availability, data authentication, data freshness, self-organization, time synchronization, and secure localization are considered while implementing security mechanisms in WSN.

A WSN consists of several individual nodes that are capable of sensing physical parameters. Recently, WSNs are widely used for developing real-time monitoring systems. The configuration of a typical wireless sensor network is illustrated in figure 9.1.

Wireless Sensor Networks are used for the sensing, processing, analysis, storage, and mining of data in an environment. The applications of WSN are in:

1. Internet of Things (IoT).
2. Surveillance and Monitoring for security, threat detection.
3. Monitoring of environmental temperature, humidity, and air pressure.
4. Noise Level of the surrounding.
5. Medical applications like patient monitoring and detection of emergencies.
6. Agriculture to determine the optical level of irrigation, use of fertilizers, and pesticides.
7. Landslide Detection.

The major components of wireless sensor networks are:

- Sensor Nodes: They are used to capture the environmental variables and which is used for data acquisition. Sensor signals are converted into electrical signals.
- Radio Nodes: It is used to receive the data produced by the Sensors and sends it to the WLAN access point. It consists of a microcontroller, transceiver, external memory, and power source.
- Wireless LAN Access Point: It receives the data which is sent by the Radio nodes wirelessly, generally through the internet.

- Evaluation Software: The data received by the WLAN Access Point is processed by a software called as Evaluation Software for presenting the report to the users for further processing of the data which can be used for processing, analysis, storage, and mining of the data.

The major merits of WSNs are:

1. Low cost: WSNs consist of small, low-cost sensors that are easy to deploy, making them a cost-effective solution for many applications.
2. Wireless communication: WSNs eliminate the need for wired connections, which can be costly and difficult to deploy and maintain. Wireless communication also enables flexible deployment and reconfiguration of the network.
3. Energy efficiency: WSNs use low-power devices and protocols to conserve energy, enabling long-term operation without the need for frequent battery replacements.
4. Scalability: WSNs can be scaled up or down easily by adding or removing sensors, making them suitable for a range of applications and environments.
5. Real-time monitoring: WSNs enable real-time monitoring of physical phenomena in the environment, providing timely information for decision-making and control.

The main demerits of Wireless Sensor Networks (WSN) are the following:

- Limited range: The range of wireless communication in WSNs is limited, which can be a challenge for large-scale deployments or environments with obstacles that obstruct radio signals.
- Limited processing power: WSNs use low-power devices, which may have limited processing power and memory, making it difficult to perform complex computations or support advanced applications.
- Data security: WSNs are vulnerable to security threats, such as eavesdropping, tampering, and denial of service attacks, which can compromise the confidentiality, integrity, and availability of data.
- Interference: Wireless communication in WSNs can be susceptible to interference from other wireless devices or radio signals, which can degrade the quality
- Deployment challenges: Deploying WSNs can be challenging due to the need for proper sensor placement, power management, and network configuration, which can require significant time and resources.

The challenges faced by a modern wireless sensor networks are the following:

1. *Limited power and energy:* WSNs are composed of battery-powered sensors that have limited energy resources. This makes it challenging to ensure that the network can function for long durations of time without the need for frequent battery replacements.

2. *Limited processing and storage capabilities:* Sensor nodes in a WSN are very small and have limited processing and storage capabilities. This makes it difficult to perform complex tasks or store large amounts of data.
3. *Heterogeneity:* WSNs often consist of a variety of different sensor types and nodes with different capabilities. This makes it challenging to ensure that the network can function effectively and efficiently.
4. *Security Issues:* WSNs are vulnerable to various types of attacks, such as eavesdropping, jamming, and spoofing. Ensuring the security of the network and the data it collects is a major challenge.
5. *Scalability:* Wireless Sensor Networks often need to be able to support a large number of sensor nodes and handle large amounts of data. Ensuring that the network can scale to meet these demands is a significant challenge.
6. *Interference and Its Mitigation:* WSNs are often deployed in environments where there is a lot of interference from other wireless devices. This can make it difficult to ensure reliable communication between sensor nodes.
7. *Reliability:* WSNs are often used in critical applications, such as monitoring the environment or controlling industrial processes. Ensuring that the network is reliable and able to function correctly in all conditions is a major challenge.

9.9.1   WIRELESS ADHOC NETWORK

Wireless Adhoc Network is a wireless network deployed without any framework or infrastructure. They include wireless mesh networks, mobile ad-hoc networks, and vehicular ad-hoc networks. The origin of Wireless Adhoc Networks can be traced to the Defense Advanced Research Project Agency (DARPA) and Packet Radio Networks (PRNET) which evolved into the Survival Adaptive Radio Networks (SARNET) program. Wireless ad-hoc networks, in particular mobile ad-hoc networks (MANET), are growing very fast as they make communication simpler and progressively accessible. In any case, their conventions or protocols will in general be hard to structure due to topology dependent behavior of wireless communication, and their distribution and adaptive operations to topology dynamism. They are allowed to move self-assertively at any time. So, the network topology of MANET may change randomly and rapidly at unpredictable times. This makes routing difficult because the topology is continually changing and nodes cannot be expected to have steady data storage. Typical applications of Wireless Adhoc Networks are:

- Data Mining.
- Military Battlefield applications.
- Commercial sector applications.
- Personal Area Network (PAN) or Bluetooth.

9.10   INTERNET OF THINGS (IOT)

The Internet of things (IoT) is a global Internet-based technical architecture facilitating the exchange of goods and services in global supply chain networks, that has

Table 9.1

Global Status of Non-IoT & IoT Devices in Billions

Year	Devices Under Non-IoT Category	Devices Under IoT
2022	10.10	16.40
2023	10.20	19.80
2024	10.21‡	24.44‡
2025	10.30‡	30.90‡

‡ shows projections.

an impact on the security and privacy of the involved stakeholders. The term Internet of Things was first used by Kevin Ashton in 1999 while working in the field of networked RFID (radio frequency identification) and emerging sensing technologies. However, practical implementations were reported only during the years 2008-2009. The growing number of physical objects that are being connected to the Internet at an unprecedented rate, paved the way to realizing the idea of the Internet of Things (IoT). A basic example of such objects includes thermostats and HVAC (Heating, Ventilation, and Air Conditioning) monitoring and control systems that enable *smart homes* to *smart cities*. There are also other domains and environments in which the IoT can play a crucial role and improve the quality of our lives. These applications include transportation, health care, industrial automation, and emergency response to natural and man-made disasters where human decision-making is difficult.

Note that in 2010, the number of everyday physical objects and devices connected to the Internet was around 12.5 billion. In 2023, there were nearly 25 billiom devices connected to the IoT. In essence, the IoT consists of the following elements: *identification, sensing, communication, computation, services, and semantics-the ability the ability to extract knowledge smartly by different machines to provide the required services.* The global status of connected devices to IoT and non-connected devices is shown in Table 9.1.

The total number of connected IoT devices is expected to reach 125 billion by 2030[3]. It is also interesting to note that global data transmissions are expected to increase from 20-25% annually to 50% per year, on average, in the next 15 years.

The IoT is enabled by the latest developments in RFID, smart sensors, communication technologies, and Internet protocols. The basic concept of IoT is to have smart sensors collaborate directly without human involvement to deliver a new class of applications. The current revolution in Internet, mobile and machine-to-machine (M2M) communication technologies can be seen as the first phase of the IoT. In the coming years, the IoT is expected to bridge diverse technologies to enable new applications by connecting physical objects together in support of intelligent decision

---

[3] As per a recent analysis by IHS Markit (Nasdaq: INFO), the number of devices connected to Internet of Things (IoT) worldwide will jump 12% on average annually; from nearly 27 billion in 2017 to 125 billion in 2030

making. The proliferation of devices connected to the Internet, and the increase in what is known as *Cyber-Physical Systems (CPS)* lead to the concept of *Big Data* analytics .

The IoT enables physical objects to see, hear, think, and perform jobs; by having them talk to one another, to share information, and to coordinate decisions. The IoT transforms these objects from being traditional to *smart devices* by exploiting its underlying technologies such as ubiquitous and pervasive computing, embedded devices, communication technologies, sensor networks, Internet protocols, and applications. The typical applications of IoT include the following:

1. Transportation-Tracking and Monitoring.
2. Smart Health care.
3. Agriculture.
4. Smart Homes and Smart Cities.
5. Smart Vehicles.
6. Schools.
7. Markets.
8. Industries.

The communication technologies used in IoT are RFID, Near-Field Communication (NFC), WiFi, Bluetooth, IEEE 802.15.4, Z-wave, Ultra-Wide Bandwidth (UWB), and LTE-Advanced. The main challenges faced by IoT now are the following:

1. Security Concerns: IoT devices are often vulnerable to security breaches, leading to data leaks, privacy violations, and even physical harm if connected to critical infrastructure. Ensuring robust security measures, such as encryption, authentication, and regular updates, is crucial.
2. Interoperability: Many IoT devices and platforms use different protocols and standards, making it difficult for them to communicate and work together seamlessly. Interoperability issues can hinder the scalability and efficiency of IoT deployments.
3. Scalability: Managing a large number of IoT devices across different locations can be challenging. Scaling up IoT deployments while maintaining performance, reliability, and cost effectiveness requires careful planning and infrastructure.
4. Data Management: IoT generates vast amounts of data, often in real-time, which can overwhelm traditional storage and processing systems. Efficiently managing, analyzing, and deriving actionable insights from IoT data is a significant challenge.
5. Privacy Concerns: IoT devices collect a wealth of personal and sensitive data, raising concerns about privacy and consent. Ensuring compliance with data protection regulations and implementing privacy-preserving technologies are essential for building trust among users.
6. Reliability and Maintenance: IoT devices deployed in remote or harsh environments may experience reliability issues due to factors like connectivity disruptions, hardware failures, or environmental conditions. Regular maintenance and remote diagnostics are necessary to ensure optimal performance and uptime.

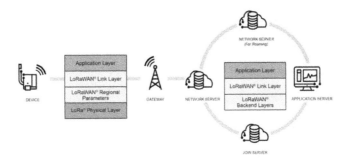

Figure 9.2 The Network Architecture of LoRaWAN

7. Cost: The cost of IoT hardware, software, connectivity, and infrastructure can be significant, especially for large-scale deployments. Finding a balance between costs and benefits while ensuring ROI is essential for the success of IoT initiatives.

9.10.1 LORAWAN TECHNOLOGY

The LoRaWAN is a Low Power, Wide Area Network (LPWAN) protocol designed to wirelessly connect battery-operated *things (devices)* to the internet in regional, national, or global networks, and targets key Internet of Things (IoT) requirements such as bi-directional communication, end-to-end security, mobility and localization services. The full form of LoRaWAN is *Long Range Wide Area Network*. The LoRaWAN specification is developed and maintained by the LoRa Alliance; an open association of collaborating members. LoRaWAN architecture is deployed in a *star-of-stars topology* in which gateways relay messages between end-devices and a central network server. The gateways are connected to the network server via standard IP connections and act as a transparent bridge, simply converting RF packets to IP packets and vice versa. The wireless communication takes advantage of the Long Range characteristics of the LoRa physical layer, allowing a single-hop link between the end-device and one or many gateways. All modes are capable of bi-directional communication, and there is support for multicast addressing groups to make efficient use of spectrum during tasks such as Firmware Over-The-Air (FOTA) upgrades or other mass distribution messages. The LoRaWAN architecture is shown in figure 9.2.

LoRaWAN has a wide-area coverage with a distance record of 766 km. While that may be its maximum standard range, a typical range in practice is 5 km in an urban environment and 10 km in rural line-of-sight conditions. LoRaWAN vertical markets include the following:

1. Smart Agriculture.
2. Smart Buildings.
3. Smart Cities.
4. Smart Industries.
5. Smart Logistics.
6. Smart Utilities.

9.10.2   WIFI HALOW

WiFi HaLow is a wireless protocol from the WiFi Alliance. This technology leverages the 900 MHz band, allowing for longer-range connections and lower power consumption compared to traditional Wi-Fi. It is a promising innovation, particularly for IoT devices. Newracom^{TM} exhibited the capabilities of WiFi HaLow in the year 2023, with a live demonstration that included cameras, sensors, and smart shelf labels that were operated through a single WiFi HaLow access point. The different devices showcased the wide range of WiFi HaLow IoT applications. While both WiFi HaLow and LoRaWAN are in the LPWAN category, WiFi HaLow will continue to grow in use and be deployed where other LPWAN technologies fall short. WiFi HaLow is an IP-based network and an open standard based on IEEE 802.11ah and its network implementation is built upon the same architecture as WiFi.

9.10.2.1   Application Range

WiFi HaLow is marketed as having a range of 1 km; however, testing revealed an even higher observed capability. In line-of-sight tests, the range was 3 km for data communication and over 600m for HD-quality video transmission with WiFi HaLow. WiFi HaLow offers different channel configurations and modulation techniques that can maximize range or data speeds, and even more interesting, it can also be deployed in a mesh topography for an even further range if needed.

9.10.2.2   Data Throughput

WiFi HaLow data rates range from 150 kbps to over 15 Mbps, over 600 times faster than LoRaWAN while still achieving good range. In open environments, WiFi HaLow has demonstrated over a 1 Mbps throughput at a distance of 3 km. Whereas LoRaWAN can only truly handle a network of sensors, WiFi HaLow can support bidirectional communication applications ranging from sensors to wireless HD video. This has opened up several new ideas and solutions for large-area IoT deployments. Commercial buildings and Industrial IoT applications have been early adopters of WiFi HaLow for use cases requiring higher throughput than what LoRaWAN can provide.

9.10.2.3   Battery Life

If we calculate energy efficiency by measuring Bits per Joule, we can see that WiFi HaLow is over 7 times as energy efficient as LoRaWAN in Transmitter (Tx)/Receiver (Rx) and yields a 50% increase in battery life. The WiFi HaLow device would complete transmission and go to sleep while the LoRaWAN device would still be transmitting.

9.10.3   RADIO FREQUENCY IDENTIFICATION (RFID)

Radio Frequency IDentification (RFID) is the first technology used to realize the Machine-to-Machine (M2M) concept (RFID tag and reader). The RFID tag represents a simple chip or label attached to provide the identity of objects. The RFID reader transmits a query signal to the tag and receives reflected signal from the tag, which in turn is passed to the database. The database connects to a processing center to identify objects based on the reflected signals within a (10cm to 200m) range. RFID tags can be active, passive, or semi-passive/active. Active tags are powered by battery while passive ones do not need battery. Semi-passive/active tags use on-board power only when needed.

The most critical component of an RFID system is the tag. There are both passive and active RFID tags, which operate in several different frequencies (high frequency, low frequency, etc.). A passive tag is energized by the signal from the RFID antenna, while an active RFID tag emits its own stronger signal. To ensure that the readability is acceptable for a particular use case, it is very vital to choose and test the RFID tags. RFID printers allow the user to generate tags on demand within the customer premises, or tags can be bulk pre-ordered from vendors. After the tag, the RFID reader and antenna is critical. Antennas *interrogate* tags which are within range in order to extract information, and the RFID reader controls connected antennas and provides filtering of results, which are then presented to the RFID software. The following are the main applications of the RFID technology.

1. Asset Tracking and Management: RFID technology is extensively used for asset tracking, allowing businesses to monitor and manage their valuable assets efficiently. Asset tracking applications may require active RFID tags, but with the right software, it is possible to achieve the same benefits with a passive RFID tag.
2. Inventory Management: RFID is an enabler of real-time inventory tracking. Improved inventory control and accuracy has a broad range of benefits, from fewer stock outs and lost orders to reduced manual labor and improved utilization of company cash. A well-implemented RFID application for inventory management can eliminate the need for staff to manually scan or otherwise track in the receipt, shipment, and movement of materials.
3. Supply Chain Management: RFID tracking enables end-to-end visibility of goods in the supply chain, reducing the risk of lost or stolen items and improving overall efficiency.
4. Quality Control: Active or passive RFID tags can be used to track the progress of products through the production process, enabling quick identification of any quality issues.
5. Cross-docking with RFID: To be efficient, cross-docking operations must move fast enough. RFID technology can be used to automate the process of sorting and redistributing incoming goods, reducing manual labor and increasing speed. An RFID antenna configuration at dock doors can automatically detect the arrival and departure of inventory.
6. Tracking of Work-in-Process: RFID can capture the arrival and departure of items at critical points in the process, and can document that correct operations were performed.

7. Authentication of products with RFID: A passive RFID tag is inserted into the item can allow the buyer of a product to verify that they are not accepting a fake item.

8. Maintenance Management with RFID: The components are labeled with passive RFID tags can help the operator get to the right spot, and background tracking via software can validate that maintenance was performed. Tags can be generated on-demand with RFID printers, and can be used to track the maintenance and repair history of assets, helping to improve the overall reliability and lifespan of equipment.

9. Safety Management and RFID: An RFID tag can be used to track the location of hazardous materials, helping the safety department personnel to ensure that they are stored and used safely.

10. Warehouse Management with RFID: Using RFID technology, we can automate the picking and packing of orders, reducing manual labor and increasing speed, efficiency, and accuracy.

9.10.4   CYBER-PHYSICAL SYSTEM (CPS)

A Cyber-Physical System (CPS) is a system that integrates *physical* and *computational components* to monitor and control the physical processes in a seamless manner. In other words, A cyber-physical system consists of a collection of computing devices communicating with one another and interacting with the physical world via sensors and actuators in a feedback loop. The CPS combines the sensing, actuation, computation, and communication capabilities, and leverage these to improve the overall performance, safety, and reliability of the physical systems. Examples for Cyber-Physical Systems are autonomous (self-driving) cars, and unmanned aerial vehicles (UAV) or drones. The key features of CPS are *reactive computation, concurrency, feedback control of the Physical World, Real-Time computation, and safety-critical application.* Typical use cases of Cyber-Physical Systems are:

- Industrial Control Cyber-Physical Systems (CPS).
- Smart Grid CPS.
- Medical CPS.
- Smart Vehicles/Automotive CPS.
- Household CPS.
- Aerospace CPS.
- Military CPS.

9.10.4.1   Challenges of Cyber-Physical Systems

The main challenges in developing industrial cyber-physical systems (CPS) include the need for rigorous numerical computation, limitations of current wireless communication technology with regard to bandwidth, computational, and storage limitations due to mobility and energy consumption, and the increased complexity of industrial systems. Another major concern is data security and data privacy.

9.10.5 BIG DATA ANALYTICS

Big Data Analytics plays a pivotal role in optimizing Cyber-Physical Systems (CPS) and IoT deployments. By processing vast streams of data generated by interconnected devices, it enables real-time insights, predictive maintenance, and efficient resource allocation. This empowers industries to enhance productivity, safety, and decision-making in an increasingly interconnected world.

In general, the term "big data" refers to datasets that are so too large and complex that traditional data processing applications are inadequate to deal with them efficiently. While there is no specific threshold that determines when data becomes *big data,* it typically involves datasets that are too large to be processed using conventional database management tools or methods. Generally, big data is characterized by the **four Vs**: volume, velocity, variety, and veracity. We will now examine the above 4 characteristics of big data.

1. Volume: This refers to the sheer amount of data being generated, collected, and stored. Big data often involves terabytes ($1TB = 10^{12}B$), petabytes ($1PB = 10^{15}B$), or even exabytes ($1EB = 10^{18}B$) of data.
2. Velocity: This refers to the speed at which data is generated and processed. With the proliferation of sensors, social media, and other sources, data is being generated at an unprecedented rate, requiring real-time or near-real-time processing.
3. Variety: This refers to the diverse types of data being generated, including structured data (like traditional databases), semi-structured data (like XML or JSON files), and unstructured data (like text, images, videos, etc.).
4. Veracity: Veracity refers to the accuracy and reliability of the data, its and value (the potential insights or value that can be extracted from the data).

9.11 CONCLUDING REMARKS

In this chapter, we discussed various advanced communication systems and protocols. We began with various multiple access schemes like SDMA, TDMA, FDMA, and CDMA and their spectral efficiencies. OFDM was discussed then. Various random access schemes and error control coding were discussed. Software-defined radio networks and Cognitive radio networks were treated in detail. We also considered the recent developments in satellite communication. We then considered internet of things (IoT), LoRaWAN, and WiFi HaLow protocols. Finally, we discussed RFID, CPS, and Big Data Analytics in detail. In short, advanced communication systems represent the driving force behind modern wireless connectivity, revolutionizing how we exchange information globally. From 5G networks enabling lightning-fast data transfer to satellite communication ensuring connectivity in remote areas, these systems are shaping the future of technology. However, with great power comes great responsibility; challenges such as privacy concerns, cybersecurity threats, and digital divides must be addressed. As we continue to innovate and explore new vistas of technology, it's imperative to prioritize ethical and inclusive practices, ensuring that the benefits of advanced communication systems are accessible to all; while safeguarding against potential risks. The journey towards a connected world is ongoing, filled with both promises and challenges.

FURTHER READING

1. Vijay K. Garg, *Wireless Communications and Networking,* Morgan Kaufmann Publishers, 2007.
2. Theodore S. Rappaport, *Wireless Communication: Principles and Practice,* 2nd Edition, Pearson Education, 2022.
3. Gagandeep Kaur and Vikesh Raj, Multirate Digital Signal Processing for Software Defined Radio (SDR) Technology, *Proceedings of the First International Conference on Emerging Trends in Engineering and Technology*, IEEE Computer Society, 2008, DOI: 10.1109/ICETET/2008.207.
4. Mikio Hasegawa, et al., Optimization for Centralized and Decentralized Cognitive Radio Networks, *Proceedings of the IEEE Transactions on Signal Processing,* Vol. 102, No. 4, April 2014.
5. Hayder Al-Hraishaw, et al., A Survey on Nongeostationary Satellite Systems: The Communication Perspective, *IEEE Communications Surveys & Tutorials,* Vol. 25, No. 1, First Quarter 2023.
6. Antony Judice, Joel Livin, and Kanagaraj Venusamy, Research Trends, Challenges, Future Prospects of Satellite Communications, *Proceedings of 2nd International Conference on Advance Computing and Innovative Technologies in Engineering (ICACITE)*, 2022, pp. 1140-43, DOI: 10.1109/ICACITE53722.2022. 9823531.
7. Maneesha V. Ramesh, Aswathy B. Raj, and Hemalatha T., Wireless Sensor Network Security: Real-Time Detection and Prevention of Attacks, *Proceedings of 2012 Fourth International Conference on Computational Intelligence and Communication Networks (CICN)*, IEEE Computer Society, November 2012, DOI: 10.1109/CICN.2012.209.
8. K. Deepa Thilak, et al., Secure Data Transmission in Wireless Adhoc Networks by Detecting Wormhole Attack, *Proceedings of the 8th International Conference on Smart Structures and Systems (ICSSS)*, 2022, DOI: https://doi.org/ 10.1109/ICSSS54381.2022.9782298.
9. Ala Al-Faqaha, et al., Internet of Things: A Survey on Enabling Technologies, Protocols, and Applications, *IEEE Communication Surveys & Tutorials,* Vol. 17, No. 4, Fourth Quarter 2015, DOI: 10.1109/COMST.2015.2444095.
10. Yongming Huang, et al., Signal Processing for MIMO-NOMA: Present and Future Challenges, *IEEE Wireless Communications*, Vol.25, No. 2, April 2018.
11. Joseph Mitola III and Gerald Q. Maguire, Jr., Cognitive Radio: Making Software Radios More Personal, *IEEE Personal Communications*, Vol.6, No. 4, August 1999. DOI: 10.1109/98.788210.

EXERCISES

1. In a wireless communication network employing SDMA, there are 4 users sharing the same frequency band. If each user requires a minimum of 10 MHz bandwidth

and the available spectrum is 50 MHz, what is the maximum number of users that can be accommodated simultaneously?

2. In a TDMA system, there are 8 users sharing the same time slot. If each user is allocated 20 milliseconds for transmission, what is the total frame duration?

3. In an FDMA system, the available bandwidth is 120 kHz, and each user requires 10 kHz bandwidth for transmission. There is a guard band of 2kHz per user. How many users can be accommodated simultaneously?

4. An OFDM system employs 64 subcarriers spaced 15 kHz apart. If the symbol duration is 4 microseconds, what is the total bandwidth occupied by the OFDM signal?

5. A (7, 4) linear block code is used for error detection and correction. How many data bits can be transmitted using this code, and how many parity bits are added for error detection and correction?

6. Consider a convolutional code with a constraint length of k=3 and a code rate of 1/2. If the encoder input rate is 1 Mbps, what is the output rate of the encoder?

7. A smart home system has 20 IoT devices transmitting data packets every 5 seconds. If each packet size is 100 bytes, what is the average data rate of the system?

8. A particular company generates 10 TB of data everyday. If they want to store this data for one year, how much storage capacity do they need in petabytes (PB)?

9. A cyber-physical system controls a manufacturing process with a control loop frequency of 100 Hz. If each control message is 200 bytes in size, what is the required data rate for communication between the cyber and physical components?

10. A communication system employs TDMA, FDMA, and CDMA (Code Division Multiple Access). There are 6 users in total. Each user requires 10 kHz bandwidth for FDMA, 5 milliseconds time slot for TDMA, and 10 chips per bit for CDMA. What is the total bandwidth required for the system?

10 Machine Learning for Communication Systems

Machine Learning is a recent development in the Artificial Intelligence (AI) field, wherein computing machines are capable of *learning* based on their past experience and data sets. Machine learning focuses on the development of algorithms and statistical models that enable computers to learn from and make predictions or decisions based on data, without being explicitly programmed for each task. In essence, machine learning algorithms allow computers to identify patterns in data and make data-driven decisions or predictions.

The origin of machine learning can be traced back to the mid-20th century. While the concept of machine learning has its roots in the early development of artificial intelligence, it began to take shape as a distinct field in the 1950s and 1960s with the work of researchers such as Arthur Samuel. Arthur Samuel is credited with coining the term *machine learning*, way back in 1959. He pioneered the development of self-learning programs, particularly in the context of game-playing programs that improved their performance over time through past experience.

However, until the advent of more powerful computers and the availability of large datasets in the late 20th century, machine learning could not begin to flourish as a practical and widely applicable/accepted technology. Since then, advances in computing power, algorithms, and data availability have propelled machine learning into various applications across industries, including finance, health care, transportation, communications, and more. In short, the future of wireless communication systems entirely lies on machine learning.

10.1 INTRODUCTION

In the evolving field of communication systems, machine learning emerges as a transformative force, revolutionizing the way data is processed, transmitted, and utilized. In this chapter, we study in detail, the strong relationship between machine learning and communication systems. By harnessing the power of advanced algorithms and data analytics, machine learning techniques empower communication systems to adapt, optimize, and innovate in real-time. From enhancing spectrum efficiency to enabling intelligent routing and dynamic resource allocation, the integration of principles of machine learning in the design and implementation of communication systems, unlock unprecedented levels of performance, reliability, and scalability in communication networks. Through exploration of key concepts, methodologies, and practical applications, this chapter illuminates the pivotal role of machine learning in shaping the future of communication technologies.

DOI: 10.1201/9781003527589-10

10.2 MACHINE LEARNING (ML)

Artificial Intelligence (AI) enables computer systems to perform tasks normally requiring human intelligence and intervention. Machine learning gives computers the ability to learn without being explicitly programmed to do so. Machine learning, a subfield of artificial intelligence, focuses on the development of algorithms and statistical models that enable computers to learn and make predictions or decisions without being explicitly programmed. It involves training algorithms on large datasets to identify patterns and relationships and then using these patterns to make predictions or decisions about new data. There are mainly three classes of machine learning:

1. Supervised Learning-This method is used when there are *training data* along with the labels (tags) for the correct classification.
2. Unsupervised Learning-In this, our main objective is to find the patterns or groups in the dataset at hand, and the dataset is unlabeled.
3. Reinforcement Learning-It is based on the dynamic interaction between an agent and the external environment, which can process data online. The agent learns by exploring the environment and rewarded or penalized actions, and optimizes the cumulative reward with the best sequence of actions.

Over the past several years, the three classes of ML; viz. supervised learning, unsupervised learning, and reinforcement learning; have all been extensively studied in the field of wireless communication. As explained in chapter 9, wireless communication is playing a major role in the everyday life of the humankind. Fifth generation (5G) and the evolving 6G advanced wireless communication techniques are expected to support various upcoming, hitherto unknown services. The principles of machine learning (ML) are deemed to optimize the performance of 5G/6G wireless systems by tackling complex problems which cannot be solved using traditional mathematical models and the associated algorithms.

10.2.1 APPLICATIONS SUPPORTED BY ML

We can identify the following applications that are supported by ML: Adaptive Modulation and Coding (AMC), Channel Equalization, Channel Coding, Beamforming, Load Prediction, and Trajectory Prediction. We will consider all of them in detail.

10.2.1.1 Adaptive Modulation and Coding (AMC)

Adaptive Modulation and Coding (AMC) is a promising technology for wireless systems, which can adjust the modulation order and/or coding rate to adapt to the fading channel variation. However, present AMC implementations are either inaccurate owing to the approximations based on models; or complicated owing to the large size of lookup tables. In this context, ML can be used to optimize AMC-assisted wireless communication. To implement ML-based adaptive modulation and coding, supervised learning and reinforcement learning are widely used. The most commonly used supervised learning algorithms are k-nearest neighbor (k-NN), support vector machine (SVM), decision trees, and neural network (NN).

The fact that time-varying wireless channels and nonlinear interference may prevent the sample data from accurately representing all scenarios is a clear drawback of the supervised learning approach. On the other hand, the reinforcement learning strategy can facilitate in-person learning from the surroundings and facilitate online learning.

10.2.1.2 Channel Equalization

The main technique for reducing nonlinear distortions and inter-symbol interference (ISI) in wireless communication is channel equalization. Traditional equalizers' performance could be enhanced by ML-based equalizers since ML can accomplish adaptive signal processing. Artificial Neural Network (ANN) based channel equalizers are able to extract the essential characteristics of time-varying wireless channels and reduce inter-symbol interference. Various ANN structures like Multi-Layer Perceptrons (MLP); Functional Link Artificial NN (FLANN) (which has lower computational complexity because it has no hidden layers); Recurrent Neural Network (RNN) which has the ability to model a nonlinear filter with infinite memory, Fuzzy Neural Network (FNN), and Deep learning NN (DNN) based equalizers have good representation capability, can exhibit robustness equalization to mitigate ISI.

10.2.1.3 Channel Coding

Channel coding is used to solve the problem of imperfect wireless channels and correct errors. In recent years, Turbo codes, Low Density Parity Check (LDPC) codes and polar codes are widely used in channel coding. Recently, deep learning is extensively explored to promote the development of channel coding and provide a universal encoder or decoder. In traditional communication systems, the original information bit sequence is converted to an encoded sequence through channel encoding. This mapping can be learned by Deep-learning Neural Network (DNN), which indicates that DNN can replace the traditional decoders to achieve high accuracy and efficiency.

10.2.1.4 Beamforming

The millimeter wave (mmWave) and massive-Multiple Input Multiple Output (mMIMO) technologies are applied to overcome the scarce spectrum availability and strong attenuation. Hybrid beamforming is a feasible solution for mMIMO systems which are operating at mmWave carrier frequencies.

10.2.1.5 Load Prediction

5G advanced wireless systems use higher frequencies to meet the stringent application demands, and hence, more base stations will be deployed to cover the same area of 4G wireless systems. Energy saving has become a key issue to overcome the large energy consumption due to the dense base stations. Yet, the accuracy of traditional energy-saving strategies based on current load rather than the future load is insufficient, which may lead to Ping-Pong effect. Ping-pong effect in 5G wireless is the rapid switching of devices between adjacent cells, optimizing signal strength for seamless and uninterrupted connectivity. Therefore, load prediction is necessary to improve the performance of energy saving.

When the predicted traffic load of a certain cell is lower than the threshold, that cell can be switched off and its users can be offloaded to a new cell. Load prediction can also be used for load balancing, which aims to distribute traffic load to different network nodes to maximize the use of network resources. A cell switch scheme using ML-based load prediction can achieve energy savings.

10.2.1.6   Trajectory Prediction

A large number of small cells is required to support the dense base stations for 5G advanced wireless systems. This situation may lead to more frequent handovers. Thus, users may need to move from one cell to another cell frequently, and this will result in high latency and reduction of throughput, which will eventually affect network performance and cause user dissatisfaction. Therefore, it is vital to achieve trajectory prediction of users to benefit network management and resource allocation. ML algorithms are widely considered to improve the performance of trajectory prediction; one such algorithm is the Long Short Term Memory (LSTM). Long short-term memory (LSTM) is a Recurrent Neural Network (RNN) based architecture with multiple memory cells, which is suitable for load prediction and trajectory prediction.

Let us now examine the subtle difference between *Unsupervised Machine Learning* and *Data-Driven Machine Learning*. As mentioned earlier, unsupervised learning is a type of machine learning where the algorithm learns from unlabeled data. The goal is to find patterns or intrinsic structures in the data without explicit supervision. Clustering algorithms like k-means, hierarchical clustering, or density-based clustering are examples of unsupervised learning techniques. Moreover, Unsupervised learning is often used for tasks such as clustering, dimensionality reduction, or anomaly detection.

Data-Driven Machine Learning, on the other hand, refers to the broader concept of using data to drive the learning process, regardless of whether the learning is supervised or unsupervised. In data-driven machine learning, the emphasis is on leveraging data to train models and make predictions or decisions. It encompasses both supervised and unsupervised learning approaches, as well as semi-supervised and reinforcement learning methods.

In short, all unsupervised learning approaches can be considered data-driven, as they rely on data to uncover patterns or structures. However, not all data-driven approaches necessarily fall under the category of *unsupervised learning*.

10.3   DEEP LEARNING (DL)

Deep learning is a subset of machine learning (ML) that deals with algorithms inspired by the structure and function of the brain's neural networks. It aims to mimic the human brain's ability to process and learn from vast amounts of data. Deep learning algorithms, particularly artificial neural networks, with multiple hidden-layers (hence the term "deep"), have gained popularity due to their remarkable performance

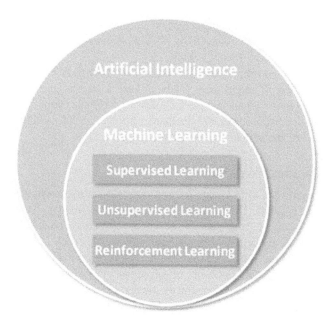

Figure 10.1   Relationship Between AI, ML, and Its Subclasses

in various tasks such as image and speech recognition, natural language processing, and game playing.

The relationship between deep learning and ML is intricate. Deep learning is a specialized technique within the broader field of ML. While traditional ML algorithms require feature extraction and selection by humans, deep learning algorithms can automatically learn hierarchical representations of data directly from raw inputs. This capability makes deep learning particularly powerful for tasks where the input data has a high level of complexity or where manually engineering features is challenging.

However, deep learning also inherits some of the challenges of traditional ML, such as the need for large amounts of labeled data for training and the risk of overfitting. Additionally, deep learning models are often computationally intensive and require substantial resources for training and inference. Despite these challenges, the advancement of deep learning has revolutionized various industries and continues to push the boundaries of what is possible in artificial intelligence (AI) and machine learning (ML). Figure 10.1 illustrates the relationship between AI, ML, and its subclasses.

10.4   MACHINE LEARNING *VERSUS* DEEP LEARNING

Machine learning and deep learning are both types of AI. Briefly speaking, machine learning is AI that can automatically adapt with minimal human interference. Deep learning is a subset of machine learning that uses artificial neural networks (ANN) to mimic the learning process of the human brain.

To gain clarity on the subtle difference between them, let us assume the following scenario. Assume that we are looking at a vast landscape filled with various objects, each with different shapes, colors, and sizes. This landscape represents the realm of data. In machine learning (ML), the observer, would carefully analyze each object, identify distinguishing features, and manually classify them into different categories. This process is akin to meticulously studying the landscape and categorizing objects based on the observations.

Having a powerful telescope that lets the observer view farther into the terrain is analogous to having deep learning (DL). Deep learning allows to feed raw data into algorithms that are modeled after the neural networks found in the human brain. Without explicit instruction, these algorithms naturally pick up on linkages, patterns, and characteristics between the items in the landscape. It appears as though the telescope recognizes objects by their intrinsic qualities and modifies itself to focus on various features of the landscape.

10.5 SPECTRUM SENSING

The cognitive radio technology is developed with the major goal of the effective use of radio spectrum resources. The spectrum sensing function plays a key role in the performance of cognitive radio networks. The increase in wireless applications and services made it essential to address the spectrum scarcity problem. A recent study by the US agency, the Federal Communications Consortium (FCC), has shown that the licensed spectrum bands are *not used* at a rate of up to 90%. In recent years, a lot of research has been done on the effective use of these spectrum bands which are either empty or are not used at full capacities. One of the notable concepts in the researche is the cognitive radio concept, introduced by Joseph Mitola III in 1999.

Cognitive radio is a software-based technology that detects the electromagnetic environment in which it operates, detects unused frequency bands, and adapts the radio's working parameters to broadcast in these bands. Spectrum sensing is a critical issue of cognitive radio technology because of the shadowing, fading, and time-varying nature of wireless channels. To sense limited or unused frequency bands, different methods for spectrum sensing have been proposed in the literature like:

1. Matched filtering.
2. Cyclostationary-based sensing.
3. Waveform-based sensing.
4. Wavelet-based sensing.
5. Eigenvalue-based sensing.
6. Energy detection sensing.

Matched filtering detection methods with shorter detection periods are preferred if certain signal information is known, such as band-width, operating frequency, modulation type and grade, pulse shape, and frame structure of the primary user (PU).

The detection performance of this method largely depends on the response of the channel. To overcome this, it requires perfect timing and synchronization in both physical (PHY) and medium access control (MAC) layers. This situation increases the complexity of calculation.

Cyclostationary detection is a method for detecting primary user (PU) transmissions by exploiting the cyclostationarity features of the received signals. It exploits the periodicity in the received primary signal to identify the presence of primary users. In this way, the detector can distinguish primary user signals, secondary user (SU) signals, or interference. However, the performance of this detection method depends on a sufficient number of samples, which increases the computational complexity.

Waveform-based spectrum sensing is only applicable to systems with known signal patterns. Such patterns include *preambles, midambles, regularly transmitted pilot patterns, and spreading sequences.* A preamble is a known sequence transmitted before each burst and a midamble is transmitted in the middle of a burst or slot. In the case of a known model, the spectrum detection function is performed by associating the received signal with a copy of itself; which is known as *auto-correlation.*

In the wavelet-based spectrum sensing method, the frequency bands of interest are usually decomposed as a train of consecutive frequency subbands. By using wavelet transform, irregularities in these bands are detected and it is decided whether the spectrum is full or empty.

Eigenvalue-based spectrum sensing does not require much *apriori* knowledge about the primary user signals and noise power. In this, the decision threshold has been obtained based on random matrix theory to make a hypothesis testing. In order to determine the presence or absence of the primary user signal, the decision threshold is compared with the test statistic formed using the ratio of the maximum or average eigenvalue to the minimum eigenvalue. Nevertheless, having a high operational complexity is a disadvantage of this method. Also, if the information of the primary users is not known precisely, energy detection-based methods with low mathematical and hardware complexities are preferred.

Energy detection is a spectrum sensing technique based on measuring the received signal energy and deciding on the presence or absence of the primary user by comparing the received energy level with a threshold. The threshold function calculation depends on the noise power. Numerous studies have been carried out to obtain the optimal threshold expression and to improve spectrum sensing performance.

The energy detection method is widely used for its simplicity in calculation and ease of application. However, the spectrum sensing performance of the energy detector is severely affected by destructive channel effects such as shadowing and fading, and noise. To minimize the negative effects caused by noise uncertainty and communication channel, the cooperative spectrum sensing model is used.

In recent years, hybrid models in which two or more detection schemes based on AI and ML are used together, have been developed to improve the performance of spectrum sensing in a cognitive radio network.

10.5.1   IMPACTS OF SPECTRUM SENSING IN 5G/6G COMMUNICATIONS

One can identify the following as the main impacts of spectrum sensing (SS):

- Efficient Spectrum Utilization: SS enables dynamic spectrum access, allowing wireless devices to utilize available spectrum bands more efficiently. This is crucial in the context of 5G and 6G, where there is a need for higher data rates and lower latency.
- Improved Spectrum Management: With SS, networks can identify unused or underutilized spectrum bands in real time, enabling better spectrum management and allocation.
- Enhanced Spectrum Sharing: SS facilitates spectrum sharing among different users and services, including licensed/primary (PU) and unlicensed/secondary users (SU), improving overall spectral efficiency.
- Cognitive Radio Networks: SS is a fundamental component in the implementation of cognitive radio networks, where devices adapt their transmission parameters based on real-time spectrum sensing results, leading to more flexible and adaptive communication systems.

The merits of spectrum sensing are:

1. Dynamic Spectrum Access: SS enables opportunistic access to spectrum bands, leading to improved spectral efficiency and increased network capacity.
2. Improved Quality of Service (QoS): By dynamically selecting the best available spectrum, SS helps maintain QoS requirements, such as latency, throughput, and reliability, especially in dense and heterogeneous networks.
3. Spectrum Efficiency: SS allows for the efficient utilization of spectrum resources by identifying unused or underutilized bands, thereby reducing spectrum wastage.
4. Enhanced Security: Spectrum sensing can also contribute to enhanced security by detecting and mitigating interference or malicious activities in the wireless spectrum.

As in any other case, the spectrum sensing is also not free from several technological and other regulatory challenges. They are the following:

- Detection Reliability: Spectrum sensing algorithms must reliably detect primary users (PU) a.k.a. licensed users, and accurately identify available spectrum opportunities, even in dynamic and noisy environments.
- Spectrum Mobility: In mobile scenarios, where users move across different geographical locations, spectrum sensing algorithms must adapt to varying propagation conditions and interference levels, posing a challenge for reliable spectrum detection.
- Coexistence and Interference: SS must address the challenges of coexistence and interference management, especially in shared spectrum environments where multiple users and services operate concurrently.

- Energy Efficiency: Spectrum sensing involves energy-intensive operations, which can drain the battery life of mobile devices. Balancing the trade-off between sensing accuracy and energy consumption is a critical challenge.
- Regulatory Constraints: Regulatory constraints and policies regarding spectrum access and sharing pose challenges for the deployment of spectrum sensing techniques, particularly in terms of spectrum sensing duration, sensing frequency, and interference mitigation strategies.

We will now discuss the automatic modulation recognition (AMR) in 5G/6G wireless communications which is a promising technological development in recent times.

10.6 AUTOMATIC MODULATION RECOGNITION (AMR)

As an intermediate step between signal detection and demodulation, automatic modulation recognition (AMR) is commonly used in cognitive radio networks to identify different types of modulation schemes used at the transmitted side. Thanks to the popularity of the Internet of Things (IoT) and related communication technologies, automatic modulation recognition (AMR) is used in various domains, from general-purpose wireless communications to military applications. Automatic modulation recognition provides essential modulation information of the incoming radio signals. AMR plays a major role in various scenarios including cognitive radio, spectrum sensing, signal surveillance, interference identification, and so on. Automatic modulation recognition is a crucial module of cognitive radio. AMR stands as a pivotal technology within the landscape of modern wireless communication systems, particularly in the context of the advanced standards like 5G and the evolving 6G.

The automatic recognition of signal modulation is the foundation of multi-mode software radio receiver (Cognitive Radio/Software Defined Radio) in civilian wireless communication. In military applications, the recognition of signal modulation is one of the key technology enablers of electronic counter measures. AMR does not need any prior information, which extracts feature values directly from acquired signals and recognizes different modulation modes.

AMR can be classified into mainly two categories: *decision theory based method* and *statistical pattern recognition*. Although the statistical mode method is easy to implement, the test results are poor under low signal-noise-rate (SNR). In contrast, the correct recognition rate of the decision theory based method for AMR is higher under low SNR. In Ref.[6], a novel multi-dimensional feature extraction of envelope spectrum characteristics and characteristics of higher-order cumulants of the modulated signal under investigation, and a new key feature is proposed to improve the performance specially in low signal-to-noise ratio (SNR). It is based on decision tree theory, which is a general method for different types of band-limited Gaussian noise modulation types. In particular, by combining the instantaneous statistic feature and high-order cumulants feature, the key features are extracted to realize the blind recognition of a total of 13 analog and digital signals-viz. AM, DSB, LSB, USB, FM, 2ASK, 4ASK, 2FSK, 4FSK, 2PSK, 4PSK, 8PSK, and $\pi/4$-DQPSK. In addition, a new characteristic parameter AT is proposed to improve the performance of modulation recognition under low signal-to-noise ratio (SNR). The simulation results show that, for all the analyzed signals, when the SNR reaches 3dB, the recognition success rate can reach more than 95%, reflecting the superiority of the method.

10.6.1 TECHNOLOGY ENABLERS FOR AMR

Machine Learning (ML), Deep Learning (DL), and Feature Extraction are the main enabling technologies for AMR. We will discuss them in detail in the following sections.

10.6.1.1 Machine Learning for AMR

Supervised learning models, such as Support Vector Machines (SVMs), Neural Networks (NNs), and Random Forests, are trained on labeled datasets to recognize modulation patterns based on features extracted from received signals. Unsupervised learning algorithms, like clustering, can also be utilized to discover inherent patterns in signal data for modulation classification.

10.6.1.2 Deep Learning for AMR

Deep learning is a subclass of machine learning. Convolutional Neural Networks (CNNs), Recurrent Neural Networks (RNNs), and their variants have been employed for modulation classification tasks, achieving state-of-the-art results. DL models can learn complex signal representations directly from raw waveform samples, eliminating the need for handcrafted features.

10.6.1.3 Feature Extraction for AMR

Feature extraction techniques, such as Fast Fourier Transform (FFT), Wavelet Transform (WT), and Cyclostationary Feature Analysis, are essential for transforming raw signal data into meaningful representations suitable for classification by ML/DL algorithms. These techniques capture relevant characteristics of modulation schemes, including frequency content, phase shifts, and temporal patterns, facilitating accurate classification.

The advantages offered by AMR include:

1. Enhanced Spectrum Efficiency: Accurate AMR enables adaptive modulation and coding schemes, optimizing spectral efficiency by selecting the most suitable modulation scheme based on channel conditions and quality requirements. This results in higher data rates and improved network capacity.
2. Robustness to Signal Variability: ML/DL-based AMR techniques exhibit robustness to signal distortions, noise, and fading effects, ensuring reliable modulation classification in dynamic and challenging environments. This enhances the resilience of wireless communication systems to interference and adverse propagation conditions.

3. Flexibility in System Design: AMR enables dynamic reconfiguration of communication systems by adapting modulation schemes based on real-time channel feedback and user requirements. This flexibility helps in efficient utilization of available spectrum resources and supports diverse applications with varying QoS constraints.

The main challenges faced by AMR now are:

- Data Complexity and Dimensionality: The high-dimensional nature of signal data poses challenges for feature extraction and ML/DL model training, requiring sophisticated techniques to capture relevant information; and at the same time reducing computational complexity and overfitting.
- Generalization and Adaptation: AMR algorithms must generalize well across different modulation schemes, signal-to-noise ratios (SNRs), and channel conditions to ensure reliable performance in diverse operating environments. Achieving robustness and adaptability to unseen channel conditions remains a formidable challenge.
- Real-Time Processing Constraints: Real-time implementation of ML/DL-based AMR algorithms pose stringent requirements on computational resources, latency, and energy consumption, requiring efficient hardware architectures and algorithmic optimizations for practical deployment.
- Security and Privacy Concerns: ML/DL models for AMR are prone to adversarial attacks and data privacy breaches, raising concerns about the security and integrity of wireless communication systems. Robustness against malicious manipulation and confidentiality of training data are critical challenges.

10.7   CONCLUDING REMARKS

In this chapter, we discussed the application of machine learning (ML) and deep learning (DL) in modern wireless communication systems. It is now established that, to meet the stringent requirements of 5G/6G wireless communication systems, we need to implement the algorithms pertaining to ML and DL. New technologies like spectrum sensing and automatic modulation recognition are based on the principles of ML and DL. It is to be noted that DL is a subclass of ML, where artificial neural networks are used to mimic the functioning (learning process) of the human brain. The software-defined radio and cognitive radio technologies have been developed with the major goal of the effective use of radio spectrum resources. Automatic modulation recognition schemes of digitally modulated signal plays a vital role in both military and civil applications like software-defined radio, electronic warfare, cognitive radio systems, radio spectrum management, threat analysis and electronic surveillance. The spectrum sensing function plays a key role in the performance of cognitive radio networks. Automatic modulation sensing is the next step in achieving the above goal. The increase in wireless applications and services resulted in a recent research and development boom in the above two technological fronts.

FURTHER READING

1. Kockaya and Develi, Spectrum Sensing in Cognitive Radio Networks: Threshold Optimization and Analysis, *EURASIP Journal on Wireless Communications and Networking*, 2020, DOI: 10.1186/s13638-020-01870-7.
2. Osvaldo Simeone, A Very Brief Introduction to Machine Learning With Applications to Communication Systems, *IEEE Transactions on Cognitive Communications and Networking,* Vol. 4, No. 4, December 2018.
3. Yingjun Zhou, et al., Applications of Machine Learning for 5G Advanced Wireless Systems, *Proceedings of the 2021 International Wireless Communication and Mobile Computing (IWCMC) Conference*, 2022, DOI: 10.1109/IWCMC51323.2021.9498754.
4. Joseph Mitola III and Gerald Q. Maguire, Jr., Cognitive Radio: Making Software Radios More Personal, *IEEE Personal Communications*, Vol.6, No. 4, August 1999. DOI: 10.1109/98.788210.
5. Peng Chu, et al., Automatic Modulation Recognition for Secondary Modulated Signals, *IEEE Wireless Communications Letters*, Vol. 10, No. 5, May 2021, DOI: 10.1109/LWC.2021.3051803.
6. Xiaodi Zhao, et al., Automatic Modulation Recognition Based on Multi-Dimensional Feature Extraction, *Proceeding of the 12th International Conference on Wireless Communications and Signal Processing*, 2020, DOI: 10.1109/WCSP49889.2020.9299797
7. Punith Kumar H.L and Lakshmi Shrinivasan, Automatic Digital Modulation Recognition Using Minimum Feature Extraction, *Proceedings of 2015 2nd International Conference on Computing for Sustainable Global Development (INDIACom)*, pp. 772–775.

EXERCISES

1. What is spectrum sensing in the context of wireless communication, and why is it essential for cognitive radio networks?
2. Discuss the different spectrum sensing techniques used in cognitive radio systems and highlight their advantages and limitations.
3. How does automatic modulation recognition (AMR) contribute to enhancing the efficiency and reliability of wireless communication systems?
4. Explain the key challenges associated with automatic modulation recognition in dynamic and noisy wireless environments.
5. What role do machine learning algorithms play in spectrum sensing and automatic modulation recognition tasks in 5G/6G systems?
6. Can you elaborate on how deep learning techniques are applied to optimize spectrum sensing and modulation recognition processes in next-generation wireless networks?
7. Discuss the potential benefits of using machine learning and deep learning models for spectrum sensing over traditional signal processing approaches.

8. How do convolutional neural networks (CNNs) contribute to automatic modulation recognition in wireless communication systems, and what are their advantages?

9. What are some of the common machine learning algorithms employed for signal classification and modulation recognition in 5G/6G networks, and how do they compare in terms of performance and complexity?

10. In what ways can machine learning and deep learning algorithms be leveraged to address the challenges of spectrum scarcity and interference mitigation in future wireless communication systems?

Index

For Product Safety Concerns and Information please contact our EU
representative GPSR@taylorandfrancis.com
Taylor & Francis Verlag GmbH, Kaufingerstraße 24, 80331 München, Germany

www.ingramcontent.com/pod-product-compliance
Ingram Content Group UK Ltd.
Pitfield, Milton Keynes, MK11 3LW, UK
UKHW021117180425
457613UK00005B/120